# BTEC First in
# Engineering

Loughborough
COLLEGE est 1909

Orders: please contact Bookpoint Ltd, 130 Milton Park, Abingdon, Oxon
OX14 4SB. Telephone: (44) 01235 827720. Fax: (44) 01235 400454.
Lines are open from 9.00 to 5.00, Monday to Saturday, with a 24 hour
message answering service. You can also order through our website
www.hoddereducation.co.uk

If you have any comments to make about this, or any of our other titles,
please send them to educationenquiries@hodder.co.uk

*British Library Cataloguing in Publication Data*
A catalogue record for this title is available from the British Library

ISBN: 978 1 444 11052 4

First Edition Published 2010
Impression number   10 9 8 7 6 5 4 3 2 1
Year                2015, 2014, 2013, 2012, 2011, 2010

Hachette UK's policy is to use papers that are natural, renewable and
recyclable products and made fromwood grown in sustainable forests.
The logging and manufacturing processes are expected to conform to
the environmental regulations of the country of origin.

Cover photo © Alistair Forrester/iStockphoto.com
Artwork by Barking Dog Art
Typeset by Fakenham Photosetting Ltd, Fakenham Norfolk
Printed in Italy for Hodder Education, An Hachette UK Company, 338
Euston Road, London NW1 3BH.

BTEC First in
# Engineering

Steve Wallis
Neil Godfrey
Anthony Carey
Michael Casey
and Anthony King

DYNAMIC
LEARNING

HODDER
EDUCATION

# Contents

# Acknowledgements

The authors and publishers would like to thank the following for use of copyrighted material in this volume:

Photo of engineer on page 1 © iStockphoto.com; Figure 1.4 © Radius Images/Alamy; Figure 1.5 © Mario Hornik/iStockphoto.com; Figure 1.6 © Scott Hirko/iStockphoto.com; Figure 1.7 © Gary Curtis/Alamy; Figure 1.17 © Mark Evans/iStockphoto.com; Figure 1.18 © Dan Moore/iStockphoto.com; Figure 1.33 © Sasha Radosavljevic/iStockphoto.com; Figure 1.35 © Chad McDermott/iStockphoto.com; Figure 1.41 © Jacob Wackerhausen/iStockphoto.com; Photo of an engineering drawing on page 45 © iStockphoto.com; Figure 2.27 © Pali Rao/iStockphoto.com; Figure 2.34 © Yong Hian Lim/iStockphoto.com; Photo of Abacus on page 85 © mark wragg/iStockphoto.com; Figure 3.1 © Cevdet Gökhan Palas/iStockphoto.com; Figure 3.2 © Alistair Forrester Shankie/iStockphoto.com; Figure 3.3 © MARIA TOUTOUDAKI/iStockphoto.com; Figure 3.4 © Don Bayley/iStockphoto.com; Figure 3.5 © Vladimir Popovic/iStockphoto.com; Figure 3.7 © Mark Evans/iStockphoto.com; Figure 3.8 © nadiya kravchenko/iStockphoto.com; Figure 3.9 © Pali Rao/iStockphoto.com; Figure 3.10 © Curventa; Courtesy of Bloodhound SSC; Figure 3.11 © PA Archive/Press Association Images; Figure 3.21 © paul kline/iStockphoto.com; Photo of atom on page 124 © Perttu Sironen/iStockphoto.com; Figure 4.1 © iStockphoto.com; Figure 4.2 © Murat Baysan/iStockphoto.com; Figure 4.3 © iStockphoto.com; Figure 4.4 © The Art Gallery Collection/Alamy; Figure 4.5 © Image Source/Rex Features; Figure 4.6 © iStockphoto.com; Figure 4.7 © Tammy Peluso/iStockphoto.com; Figure 4.8 © Sam Valtenbergs/iStockphoto.com; Figure 4.9 © Steve Rosset/iStockphoto.com; Figure 4.10 © Karl Weatherly/Getty Images; Figure 4.11 © Bob Thomas/iStockphoto.com; Figure 4.25 © Jason Lugo/iStockphoto.com; Figure 4.28 © Ian Dagnall/Alamy; Figure 4.31 © JENS SCHLUETER/AFP/Getty Images; Figure 4.32 © Perttu Sironen/iStockphoto.com; Figure 4.33 © Dejan Petkovski/iStockphoto.com; Figure 4.34 © Hugh Threlfall/Alamy; Figure 4.35 © MARIA TOUTOUDAKI/iStockphoto.com; Figure 4.36 © Andrey Volodin/iStockphoto.com; Figure 4.42 B&B Training Associates Ltd.; Figure 4.43 B&B Training Associates Ltd.; Figure 4.45 B&B Training Associates Ltd.; Figure 4.46 B&B Training Associates Ltd.; Photo of engineering materials on page 175 © Masoud Sarikhani/iStockphoto.com; Figure 8.1 © Toru Hanai/Reuters/Corbis; Figure 8.2 © Nikreates/Alamy; Figure 8.4 © Vitaliy Pakhnyushchyy – Fotolia.com; Figure 8.5 © Windsor – Fotolia.com; Figure 8.6 © Donall O Cleirigh/iStockphoto.com; Figure 8.7 © Feng Yu – Fotolia.com; Figure 8.9 © Alex Kalmbach – Fotolia.com; Figure 8.10 © Joshua Haviv/iStockphoto.com; Figure 8.12 © Sergey Shlyaev – Fotolia.com; Figure 8.16 © Guillermo Lobo/iStockphoto.com; Figure 8.19 © SHEILA TERRY/SCIENCE PHOTO LIBRARY; Figure 8.21 © Christopher Dodge – Fotolia.com; Photo of CAD drawing of a car on page 218 © ArchMen – Fotolia.com; Figure 10.1 © Siede Preis/Photodisc/Getty Images; Figure 10.2 © Lester Lefkowitz/Taxi/Getty Images; Photo of a drill on page 263 © Yuriy Nedopekin/ iStockphoto.com; Figure 14.7 © Dan Driedger/iStockphoto.com; Figure 14.8 © Alistair Forrester Shankie/iStockphoto.com; Figure 14.10 © Achim Prill/iStockphoto.com; Figure 14.14 © Josef Bosák/iStockphoto.com; Figure 14.15 © David Birkbeck/iStockphoto.com; Figure 14.16 © Lyudmil Kolev/iStockphoto.com; Figure 14.20 © michele Galli/iStockphoto.com; Figure 14.22 © Joerg Reimann/iStockphoto.com; Figure 14.28 © David Stockman/iStockphoto.com; Figure 14.36 © Dušan Zidar/iStockphoto.com; Photo of a ruler on page 309 © Andrius Maciunas/iStockphoto.com; Figure 18.14 © Andrius Maciunas/iStockphoto.com; Figure 18.15 © Will Burwell/photographersdirect.com; Figure 18.17 © bluefern – Fotolia.com; Figure 18.20 © Alan Herbert/photographersdirect.com; Figure 18.23 © Robert Down/photographersdirect.com; Figure 18.26 © Frank Wright/iStockphoto.com; Figure 18.45 © fStop/Alamy; Photo of a computer chip on page 345 © Simon Smith/iStockphoto.com; Figure 19.3 © PeteG – Fotolia.com; Figure 19.4 © Daniel Fisher/iStockphoto.com; Figure 19.7 © iStockphoto.com; Figure 19.8 © Art Directors & TRIP/Alamy; Figure 19.16 © savitch – Fotolia.com; Figure 19.17 © Nivellen77 – Fotolia.com; Figure 19.18 © iStockphoto.com; Figure 19.20 © Rob Hill – Fotolia.com.

# Introduction

You do not need to look too far to see engineering; in fact, the results and outputs of engineering are pretty much everywhere. On one end of the scale are very high-profile engineering projects, such as the new Olympic stadium, the latest F1 cars or the launch of the latest space project. At home, you may have the latest computer, games console, mobile phone or media player. On television, you may watch advertisements for the latest products without realising the latest engineering techniques that have gone into producing them. If you have been to a music festival, then the stage would have been engineered; if you have been on holiday, then the aircraft you travelled on would have been engineered. In short, if you can touch it, then it has probably involved some level of engineering. Some specialist engineering and manufacturing companies may not be well known to you, but others are household names: Ford, Honda and Ferrari; Apple, IBM, Sony and Nintendo; Sky, Orange, Vodafone and O2; Coca-Cola, Walkers and Kellogg's; Nike, Adidas and Puma.

One of the most famous engineering projects was the NASA Apollo programme that landed Neil Armstrong on the moon. Back in 1969, the NASA engineers worked to a failure rate of 99.9 per cent, meaning that from six million rocket components they expected up to 6,000 to fail or break. This was a staggeringly big risk. Today, as engineers harness the latest computer and control technology, and the latest materials, components and ingredients, they work to a failure rate approaching three in a million, or 99.9997 per cent.

Engineering does not stand still; it keeps moving, keeps developing, keeps making the limits of scientific thinking real and achievable and keeps much of the world clothed, fed and watered. But most crucially, engineering needs engineers, and this book is as good a place as any to start on that journey.

# How this book can help you

If you are reading this book, then it is highly likely you are completing the Edexcel BTEC First in Engineering as part of an extended school curriculum, on a full-time college course or as part of an engineering apprenticeship.

This book is arranged into a number of units that cover some of the most important fundamentals of engineering, technology and science.

Engineering is essentially technical problem solving and to do this you need to have knowledge, or 'know-how'. This book can help you by providing an introduction to many important topics that underpin engineering. Such topics include mathematics and science; using and selecting materials; marking out and making products using material-removal techniques; fundamental working practices; and how to understand, use and present technical information, including how to use computer-aided design packages.

**Make the grade**

This kind of box identifies activities designed to help you with your coursework and assessments. In some cases, they may point you in the direction of a particularly useful website or television programme.

# How this book is organised

The book is organised into nine units that correspond with the units you are likely to be completing at your school, college or training centre.

Each unit contains a series of sections designed to provide you with the important technical information that will help you with your coursework and assessments.

**" Team Talk**

**Team Talks are used to clarify any technical terms you may not have heard before. These are short blogs between Aisha, a student, and Steve, her teacher, that may prove useful for further learning or practical activities. "**

# About the authors

Steve Wallis, Neil Godfrey, Anthony Carey, Michael Casey and Anthony King are all lecturers at Hartlepool College of Further Education, where they have gained considerable experience delivering the Edexcel BTEC First Diploma in Engineering. The writing team are all graduate engineers and qualified teachers and, most importantly, have worked as engineers in a range of disciplines, including electrical engineering, mechanical manufacturing, and process and design engineering.

The team have previously written for Edexcel and SEMTA; Steve and Neil have additionally produced two well-received textbooks for GCSE Engineering and GCSE Manufacturing.

# Unit 1
# Working safely and effectively in engineering

# Introduction to the unit

Engineering is a challenging and enjoyable profession. During their career, engineers will inevitably be faced with the prospect of dealing with potentially hazardous situations. Many manufacturing operations and production processes cannot be carried out without some risk being involved. To reduce this risk to a minimum, there are numerous rules and regulations that must be adhered to. Some are set by statutory law; others are organisational requirements. They are all meant to ensure that work is carried out in the safest and most effective manner.

A good awareness of safe working practices is an essential item in any engineer's toolkit. Equipment and materials need to be moved from A to B around the workplace and stored appropriately. Suppose these materials were heavy, hot, toxic or flammable, or had sharp edges. How would you deal with this safely and effectively? Typical work activities in an engineering workplace might involve working with chemicals or electricity, or drilling and machining materials. How would you ensure that you did not harm yourself, your colleagues or an innocent bystander while carrying out the work?

This unit will help you to identify safe and effective methods of carrying out work activities, including handling materials and equipment. It will also help you to understand the need for cooperation and good communication when working in an engineering environment.

## Learning Outcomes

By the end of this unit you should:

- be able to apply statutory regulations and organisational safety requirements;
- be able to work efficiently and effectively in engineering.

# Grading criteria

| To achieve a pass grade you must be able to: | To achieve a merit grade you must be able to: | To achieve a distinction grade you must be able to: |
|---|---|---|
| **P1** handle materials and equipment in an engineering workplace in a safe and approved manner | **M1** carry out a risk assessment on an engineering workplace to make recommendations on the safety of materials and equipment handling, use of personal protective equipment and the potential hazards in the area | **D1** prepare a safety policy for an engineering work area including references to relevant legislation |
| **P2** select and use appropriate personal protective equipment when undertaking a given engineering activity | **M2** make recommendations for improvement of an organisation's emergency procedure | **D2** identify strengths and areas for improvement in a working relationship |
| **P3** identify hazards and risks associated with an engineering activity | **M3** identify how a work activity could be improved | |
| **P4** describe the emergency procedures to be followed in response to a given incident in an engineering workplace | | |
| **P5** prepare for and carry out an engineering work activity | | |
| **P6** maintain good working relationships with colleagues and other relevant people when carrying out an engineering work activity | | |

# Learning Outcome 1. Be able to apply statutory regulations and organisational safety requirements

## Health and Safety at Work Act 1974

Before we talk about safe methods of materials and equipment handling, it is important to gain an understanding of the laws and regulations governing our working practices. Just as traffic regulations in the United Kingdom restrict the speeds at which we can drive on the roads, workplace activities are covered by the Health and Safety at Work Act 1974. You will often see this abbreviated to HASAWA 1974.

This is an Act of Parliament that places a responsibility or duty on you, as the employee, to work safely; it also applies to your employer. It does not apply to the premises (i.e. buildings, offices or workplaces), but it does apply to the people using them.

Deaths, accidents, injuries and ill health occur in many industries, such as the agricultural, construction, manufacturing or service sectors. These are all areas where you, as an engineer, could be working.

Since the Health and Safety at Work Act 1974 was introduced, there has been a substantial reduction in the number of work-related injuries and deaths. Unfortunately, even with the HASAWA 1974 in place, there are around 200 workplace deaths, 800,000 accidents and 150,000 non-fatal injuries per year in the UK. An estimated two million people suffer from ill health caused by or made worse through work.

However, do not worry too much about the statistics. When compared with other countries, Great Britain has one of the lowest work-related fatal injury rates in Europe.

**Figure 1.1** Traffic regulations restrict speeds just as HASAWA 1974 covers workplace activities

So, what are the consequences of not adhering to the HASAWA 1974? Let us consider the traffic regulations again. If you break the speed limit on a public road, you might cause an accident or be stopped and prosecuted by the police. The police are there to enforce the traffic regulations. Even if you think you can get away with breaking the speed limit, you are breaking the law and putting your own and other people's lives at risk.

The Health and Safety at Work Act is enforced by a government agency called the Health and Safety Executive (HSE) and also by local authorities. The main aim of the HSE is to protect people against risks to health and safety arising from work activities. If you break the HASAWA 1974, you can expect to come into contact with the HSE and you could be prosecuted. The HSE has the power to take a person to a court of law, which can lead to the issue of unlimited fines and up to two years' imprisonment.

Now we know that it is not a good idea to break the Health and Safety at Work Act 1974 (or any other law!), we need to gain a better understanding of our responsibilities and duties.

The Health and Safety at Work at Act can be viewed and downloaded from the HSE website. It is a lengthy document, consisting of about 130 pages of legal jargon. There are also more concise guides available, outlining the main points of the act. The most important parts that concern us are the duties of the employer and the duties of the employee. Let us look at what this means to the employer first.

## General duties of the employer to their employees

Under the act, the employer must ensure, as far as is reasonably practicable, the health, safety and welfare at work of the employees. Note the words 'reasonably practicable'. In essence, what this means is that measures to reduce risks to people must be adopted, unless there is a disproportionate sacrifice in implementing such measures.

**66 Team Talk**

Aisha: **'Explain disproportionate and proportionate, please.'**

Steve: **'An employer would not be expected to spend £1 million on preventing slightly bruised elbows to five employees. This is grossly disproportionate. However, an employer would be expected to spend £1 million on preventing an explosion that could injure 50 employees. This is proportionate.'** 99

To comply with the HASAWA, the employer has to carry out some specific tasks to ensure your health, safety and welfare at work. These include providing a health and safety policy; providing appropriate training, instruction, information and supervision; carrying out risk assessments; and maintaining the workplace and its environment. Remember, the employer's responsibilities extend to members of the public and not just their employees.

## General duties of the employee at work

It is not just the employer who takes responsibility for the health and safety of people at work – you, as the employee, have a duty under the HASAWA 1974 too. It is worth remembering the following from the HASAWA:

*'It shall be the duty of every employee while at work –*

*to take reasonable care for the health and safety of himself and of other persons who may be affected by his acts or omissions at work.'*

**66 Team Talk**

Aisha: **'Who could the "other persons" be?'**

Steve: **'The other persons could be members of the public, colleagues, visitors, cleaners, office staff, anyone whose health, safety or welfare may be affected by something you did or did not do while going about your job.'** 99

What this means is that you must ensure you have taken all steps necessary to ensure the health and safety of yourself and other people.

We have briefly introduced the Health and Safety at Work Act 1974. The following activity requires you to investigate the HASAWA 1974 in more detail. Discuss your findings in a group before writing down your answers in full.

## Activity

1. Using the Health and Safety Executive (HSE) website (www.hse.gov.uk), find out what the general duties for employers and employees are under Sections 2, 3 and 7 of the HASAWA 1974. Make a list of at least three duties for both the employee and the employer.

2. Which section covers the duties of people concerned with premises in relation to people other than employees?

3. Would a person who designs, manufactures, imports or supplies an article for use at work have a duty or responsibility under the HASAWA 1974?

4. You witness someone at work deliberately interfering with a fire extinguisher. What would you do about this and why?

5. At work, your employer wants you to pay for a pair of safety goggles. You need these to carry out your normal job, drilling holes in a piece of material. What does the HASAWA say about this?

6. Produce a pocket-sized guide to the Health and Safety at Work Act 1974. The HSE website has a web-friendly guide, which will be helpful for this activity (http://www.hse.gov.uk/pubns/law.pdf). This could also be done as a Powerpoint® presentation or poster.

# Other important safety legislation

We have seen that, although the HASAWA is the basis of British health and safety law, it is a general or enabling act which allows the Secretary of State to make further laws without the need to pass another Act of Parliament. These are called Regulations and cover more specific health and safety matters, some of which we will look at now. However, before moving on, there are a few important terms you need to understand.

## Key word

**Guidance** – to help people understand and comply with aspects of health and safety law, the HSE often issues guidance notes. These are in the form of written or downloadable documents. They are useful for specific technical advice. In other words, if you follow the advice given in the guidance by the HSE, you can be confident that you are complying with the law.

## Activity

When reading about health and safety, you might come across the following terms:

- guidance;
- approved codes of practice (ACOP);
- regulations.

Investigate how these relate to health and safety and discuss these terms in your group. What are the differences between them?

## Other current and relevant safety regulations

Now we have an understanding of what regulations are, we can look at other important health and safety regulations. In addition to the Health and Safety at Work Act 1974, the following regulations apply across workplaces:

- **Management of Health and Safety at Work Regulations 1999:** these regulations require employers to carry out risk assessments, make arrangements to implement necessary measures, appoint competent people and arrange for appropriate information and training.
- **Workplace (Health, Safety and Welfare) Regulations 1992:** these regulations cover a wide range of basic health, safety and welfare issues, such as ventilation, heating, lighting, workstations, seating and welfare facilities.
- **Health and Safety (Display Screen Equipment) Regulations 1992:** these regulations set out requirements for work with visual display units (VDUs).
- **Personal Protective Equipment at Work Regulations 1992:** these regulations require employers to provide appropriate protective clothing and equipment for their employees.
- **Provision and Use of Work Equipment Regulations 1998:** these regulations require that equipment provided for use at work, including machinery, is safe and users are trained to use it.

## Key words

**Approved codes of practice** – an approved code of practice is useful for ensuring you comply with the requirements of a specific health and safety law. If you are prosecuted by the courts for a breach of health and safety law and you have not followed a code of practice, you are likely to be found at fault.

**Regulations** – regulations are laws, approved by parliament. Under the Health and Safety at Work Act, employers have the freedom to decide on how to control any risks inherent to work activities. There are instances, however, where the work activities are so hazardous or the control measures would be so costly that specific action must be taken to deal with them safely. In these cases, regulations are used to identify the risks and to set out the action to be taken.

- **Manual Handling Operations Regulations 1992:** these regulations cover the moving of objects by hand or bodily force.
- **Health and Safety (First Aid) Regulations 1981:** these regulations cover requirements for first aid.
- **Health and Safety Information for Employees Regulations 1989:** these regulations require employers to display a poster telling employees what they need to know about health and safety.
- **Employers' Liability (Compulsory Insurance) Act 1969:** this act requires employers to take out insurance against accidents and ill health to their employees.
- **Reporting of Injuries, Diseases and Dangerous Occurrences Regulations (RIDDOR) 1995:** these regulations require employers to report certain occupational injuries, diseases and dangerous events.
- **Noise at Work Regulations 1989:** these regulations require employers to take action to protect employees from hearing damage.
- **Electricity at Work Regulations 1989:** these regulations require people in control of electrical systems to ensure that they are safe to use and that they are maintained in a safe condition.
- **Control of Substances Hazardous to Health Regulations (COSHH) 2002:** these regulations require employers to assess the risks from hazardous substances and take appropriate precautions.
- **Lifting Operations and Lifting Equipment Regulations 1998:** these regulations aim to reduce risks to people's health and safety from lifting equipment provided for use at work.

While this list is extensive, it is by no means complete. There are many other pieces of legislation that cover specific health and safety issues in the workplace.

## Activity

Select one of the regulations from the list above. Produce a short presentation to be delivered to the rest of the group, outlining the main points associated with the regulation. You should aim to have at least six slides and for the presentation to last about ten minutes. Try to use ICT equipment, such as a computer, PowerPoint® and projector if possible. You can work in pairs if you are nervous about presenting.

Alternatively, produce a workshop poster providing information about an important health and safety regulation you know about.

# Control of Substances Hazardous to Health Regulations (COSHH) 2002

Every year, thousands of employees become ill through exposure to hazardous substances. Asthma, cancer and dermatitis are just some of the illnesses that can be caused after exposure to harmful substances. If an employee were unable to work because of this, there would be a large cost to both the employer and the employee. Costs to society can also be substantial when you take into consideration the medicines and health care required to treat the illnesses. The trained worker would have to be replaced, which would also be a cost burden for the employer.

The Control of Substances Hazardous to Health Regulations 2002 are often abbreviated to COSHH. Hazardous substances are anything that can be harmful to your health when you work with them. COSHH covers how hazardous substances are supplied, purchased, stored, used and disposed of. Nearly all workplaces will have some substances that are covered under these regulations. The purpose of COSHH is to control exposure to hazardous substances and to protect both employees and others who may be exposed through work activities.

Substances covered by the COSHH regulations are those that are used, generated or occur naturally.

Substances used directly in a work activity:

- glue;
- paint;
- cleaning products.

Substances generated from a work activity:

- welding fumes;
- soldering fumes;
- toxic fumes.

Substances that occur naturally:

- grain dust;
- blood;
- bacteria.

It is not just chemists in laboratories who will come into contact with chemicals. Many engineering work activities involve the use of chemical substances. Most are not particularly harmful if you know how to deal with them safely. Some chemicals, however, need very careful handling and, as with all chemicals, you need to know how to deal with them safely (especially if there is a spillage). To help you identify the more hazardous chemicals, a standard labelling system is used. Under a series of regulations called the Chemicals (Hazard Information and Packaging for Supply) Regulations 1994 (known as CHIP), warning signs and information about the substance and its effects are provided on labels.

If you look on the back of a bottle of household bleach, you will see a label containing a warning sign and information about the product.

## Activity

1. **Identify the warning signs shown in Figure 1.2.**
2. **Which one would be on a bottle of household bleach?**

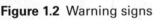

| a | b | c |

**Figure 1.2** Warning signs

By law, suppliers of dangerous chemicals must label their products with a hazard symbol, warnings, safety advice and sometimes instructions for use. A detailed information sheet, known as the 'material safety data sheet' (MSDS) or 'safety data sheet' (SDS), should be provided as a requirement of current European Union law. This is to ensure workers and the environment are protected from the effects of the hazardous chemicals.

A warning label for an industrial chemical is shown in Figure 1.3.

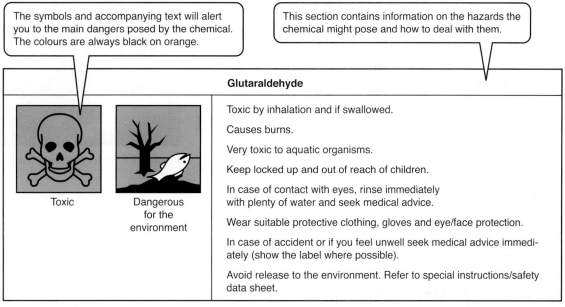

The symbols and accompanying text will alert you to the main dangers posed by the chemical. The colours are always black on orange.

This section contains information on the hazards the chemical might pose and how to deal with them.

**Glutaraldehyde**

Toxic

Dangerous for the environment

Toxic by inhalation and if swallowed.

Causes burns.

Very toxic to aquatic organisms.

Keep locked up and out of reach of children.

In case of contact with eyes, rinse immediately with plenty of water and seek medical advice.

Wear suitable protective clothing, gloves and eye/face protection.

In case of accident or if you feel unwell seek medical advice immediately (show the label where possible).

Avoid release to the environment. Refer to special instructions/safety data sheet.

**Figure 1.3** Warning label for industrial chemical

# Personal protective equipment (PPE) P1 P2 P3

Now that you are aware of some of the risks posed by hazardous substances, you need to know how to protect yourself while working with them. The Personal Protective Equipment at Work Regulations 1992 require that personal protective equipment (PPE) should be supplied by your employer for you to wear and use whenever the risks to your health and safety cannot be adequately controlled in other ways. For typical engineering work activities, you will need to wear safety boots, overalls, a hard hat and eye protection. It is your responsibility to look after them.

Some work activities require the use of special types of PPE. We will have a look at some of these. Remember that the wearing of PPE should only be necessary because other adequate hazard control measures are not practical.

As we can see from the following examples of work activities, the selection and use of the correct PPE is essential to ensure you are protected against hazards.

In the first example (Figure 1.5), a worker is using a hand-held grinder to shape a piece of metal. Table 1.1 outlines the hazards, risks and correct PPE in this situation.

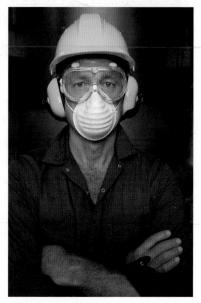

**Figure 1.4** PPE

| Hazard | Risk | PPE |
|---|---|---|
| Hot metal particles | Eye injuries | Full face visor Safety spectacles |
| | Burns to hands | Heat-proof gauntlets |
| | Burns to head | Helmet |
| Noise | Damaged hearing | Ear protection |
| Dust | Inhalation of dust | Dust mask |
| Falling objects | Broken toes | Safety boots |

**Table 1.1** Hazards, risks and PPE for a hand-held grinder

**Figure 1.5** Hand-held grinder

In the next example (Figure 1.6), showing the use of a pedestal drill, eye protection is essential. Overalls are worn with the sleeves rolled up to prevent entanglement in the rotating drill. Table 1.2 outlines the hazards, risks and correct PPE in this situation.

| Hazard | Risk | PPE |
|---|---|---|
| Metal swarf | Eye injuries | Safety goggles |
| Rotating drill | Clothing getting caught in machinery or drill | Overalls |
| Falling objects | Broken toes | Safety boots |

**Table 1.2** Hazards, risks and PPE for a pedestal drill

**Figure 1.6** Pedestal drill

## Activity

It is not good practice to wear gloves when using a pillar drill. Can you think what the hazards and risks are?

## Activity

1. **Produce a handout that shows the correct PPE to be worn for an engineering activity that you will undertake. Use the format from the previous examples to help you. Some examples of activities are given below. However, you may be given a different engineering activity.**

   - **Cutting and filing a piece of metal to shape in an engineering workshop using hand tools.**
   - **Drilling a piece of metal using a workshop pedestal drill.**
   - **Soldering electrical and electronic circuits.**
   - **Marking out a hole centre on a metal workpiece using a hammer and centre punch.**
   - **Wiring electrical circuits in a workshop.**
   - **Using workshop machine tools (for example, a lathe or milling machine) to produce an engineering component.**

2. **Using one of the activities from the list above (or one selected for you), carry out the activity using the PPE you have identified in your answer for Question 1. Obtain a witness statement or observation record to show that you have completed this activity.**

### Make the grade

This activity will help you in achieving the following grading criteria:

 **handle materials and equipment in an engineering workplace in a safe and approved manner;**

**P2** **select and use appropriate personal protective equipment when undertaking a given engineering activity;**

**P3** **identify hazards and risks associated with an engineering activity.**

# Emergency procedures

Whatever environment you are working in, you must be able to deal with emergency situations. You might have already taken part in emergency evacuation procedures at your workplace, school or college – for example, because of a fire alarm or fire drill. In industrial and commercial workplaces, similar procedures are adopted to ensure the safety of employees during emergency situations. In an emergency, personnel must be able to evacuate the workplace quickly and safely, whether that workplace is a factory, office, hotel, aircraft or engineering facility. In some cases, it may be necessary to evacuate hundreds of people. Therefore, it is important that procedures are in place so that all personnel know what to do in such a situation.

## Reporting accidents and incidents

All accidents and injuries, even near-misses, should be reported to a responsible person. This could be your supervisor, instructor or health and safety officer. It is important to report these incidents, no matter how small or trivial they may seem. Doing so may help to prevent more serious accidents or incidents occurring.

## Fire emergencies

We will now look at some emergency situations and the procedures used to deal with them. One of the most common emergencies you might have to deal with is that of a fire in your building. A few simple steps will ensure your safety and that of others around you.

1. **If you discover a fire, you must first raise the alarm to warn everyone in the vicinity of the emergency situation.** With most modern fire-alarm systems, the alarm can be sounded by pressing the button on a manual call point or a 'break glass' device. The fire alarm could be a bell, siren or klaxon, depending on the premises. If you do not warn anyone else about the fire, then there is a risk they may become trapped if the fire became serious. You may be working on machinery or equipment when you hear the fire alarm. If so, you should ensure the task you are leaving is safe before evacuating. You do not have to spend a long time doing this – it could be just switching off the machine, tightening a bolt or closing a valve.
2. **The fire service should be called.** In certain circumstances, specially trained employees can attempt to contain the fire with firefighting equipment until the fire brigade arrives. You should not attempt to tackle the fire unless you are certain you would not be putting yourself or others in danger.
3. **Evacuate the premises using the nearest fire exit.** You may have practised fire evacuations in the workplace before. In this case, you should know where to find the designated safe area outside the premises. If you are a visitor or a new employee, then you should have been informed of the emergency evacuation procedure during an induction. Make sure you close fire doors behind you to prevent the spread of smoke and fire. Never use a lift as a means of escape in a fire evacuation.

**Figure 1.7** Fire exit

Fire wardens carry out specific duties in the event of a fire emergency. They will direct you to a safe exit and advise you of the allocated assembly area. You can identify the fire warden by the high-visibility vest or armband they wear. As well as carrying out these emergency roles, they also assist on a day-to-day basis by monitoring general fire safety issues.

## Activity

1. **Produce a safety sign that explains the evacuation procedure to follow if the fire alarm sounds in an engineering workplace.**
2. **Make recommendations as to how you would deal with an incident where a colleague has a badly cut finger.**

## Make the grade

This activity will help you in achieving the following grading criterion:

**P4** describe the emergency procedures to be followed in response to a given incident in an engineering workplace.

## Activity

You have been asked to review an organisation's emergency procedure and you discover the fire evacuation plan does not include guidance for people with a sensory impairment or a disability. Describe your recommendations for improving the evacuation plan.

## Make the grade

This activity will help you in achieving the following grading criterion:

**M2** make recommendations for an improvement of an organisation's emergency procedure.

# Classification of fire types and fire extinguishers

Three elements must be present in order for a fire to start and spread. These elements can be represented by the 'fire triangle' (Figure 1.8).

If any one of the three elements is removed, then the fire cannot start or continue to burn. Oxygen usually comes from the air in the atmosphere, although sometimes oxygen is contained in pressurised containers. Heat is the ignition source. This could be a discarded match or cigarette, the Sun, a spark from a loose electrical connection or overheating electrical equipment. Fuel is the material or substance that catches fire and burns. To extinguish a fire, we need to remove one or more of the elements from the fire triangle.

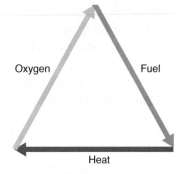

**Figure 1.8** The fire triangle

1. Remove the heat by cooling.
2. Remove the oxygen by smothering.
3. Remove the fuel by starvation.

Fire extinguishers are designed to do this for us; unfortunately, there is not a universal extinguisher for all fires. Therefore, it is very important that we choose the correct fire extinguisher if we are going to tackle a fire. There is the possibility that a fire could be made worse or the firefighter could be put at greater risk if the wrong type of extinguisher is used.

## Classification of fire types

Fires are classified into specific types under a British Standard (known as BS EN-2). These fire types are shown in Figures 1.9–1.14.

**Figure 1.9** Class A – fires involving solid materials (usually organic, such as wood, paper or textiles)

**Figure 1.10** Class B – fires involving flammable liquids and solids (including those that do and do not mix with water)

**Figure 1.11** Class C – fires involving gases (such as natural gas or propane, etc.)

**Figure 1.12** Class D – fires involving metals or powdered metals

**Figure 1.13** Class E – electrical fires. Electricity is a source of ignition and there is the risk of electrocution until it is isolated (once isolated, it can be treated as a Class A fire)

**Figure 1.14** Class F – fires involving high-temperature cooking oils (such as those found in large catering kitchens)

# Fire extinguishers

Portable fire extinguishers are predominantly red with a coloured band around them to identify the contents:

- water – red band;
- dry powder – blue band;
- foam – cream band;
- carbon dioxide – black band;
- special wet chemical – yellow band.

Fire extinguisher types and what they should and should not be used on are shown in Figures 1.15 and 1.16.

| Fire extinguisher type | Use on | Do not use on |
|---|---|---|
|  Water | Paper, wood, textiles and solid materials fires | Liquid, electrical or metal fires |
|  Powder | Liquid, electrical, wood, paper and textile fires | Metal fires |
|  AFFF foam | Liquid, paper, wood and textile fires | Electrical or metal fires |
|  Carbon dioxide ($CO_2$) — DO NOT HOLD HORN WHEN OPERATING | Liquid and electrical fires | Metal fires |

**Figure 1.15** Types of fire extinguisher

Burning cooking oils and fats are very difficult to extinguish. The very high temperatures cause the fire to reignite if conventional extinguishers are used. This is because they do not provide sufficient cooling to remove the heat from the fire triangle. A special wet chemical extinguisher works by cooling and emulsifying the burning liquid. This extinguishes the flame and also seals the surface, preventing auto-ignition.

**Figure 1.16** Wet chemical fire extinguisher

## Activity

Using the classification of fire types and the list of fire extinguishers, produce a safety sign or poster that shows the correct extinguisher to be used for each classification of fire.

Make sure you correctly identify the colour codes for each of the portable fire extinguishers in the list. Explain why halon is no longer used in new fire extinguishers.

## Hazards and risks

It is important that you understand risk assessment and risk control. This can be a complex subject; however, we will try to deal with it in a straightforward manner. Firstly, you need to understand what is meant by hazard, risk and risk control (see Team Talk).

## 66 Team Talk

Aisha: **'What is a hazard?'**
Steve: **'A live tiger! Unless it is anaesthetised, if you are near one of these, it always has the potential to cause you harm. It is a real hazard.'**
Aisha: **'What is a risk?'**
Steve: **'A tiger roaming about is a risk because it is very likely that it will cause harm to anyone near it!'**
Aisha: **'So what is risk control?'**
Steve: **'To remove the hazard, we need to remove the tiger. If we can't do that, then we avoid the hazard – don't go anywhere near the tiger! If we can't do that, then we can introduce a risk-control measure.'**
Aisha: **'What is a risk-control measure?'**
Steve: **'We'll need to put the tiger in a cage!'**

99

The example in the Team Talk illustrates a basic strategy for risk control. However, we know that this can be a complex subject when engineering systems are involved. Let's look at another example of hazard, risk and risk control before we move onto a workplace scenario.

Imagine yourself as an experienced racing driver, competing in a race in a Formula 1 car. At such high speed, the race track ahead disappears very quickly as each corner approaches. You need to react extremely quickly to keep the car on course and under control.

**Figure 1.17** Formula 1 is a high risk sport

## Activity

**What are the risks involved for you as a driver in such a situation? And what are the risks for the spectators and track marshals?**

There is certainly the risk that you might not make it round the corner and hit a crash barrier or wall, injuring yourself or others in the process. Consider what might cause the car to become out of control. It could be caused by a mistake or lack of concentration by the driver. There are external factors to consider as well; you might hit an oil patch or debris on the track, another car might make contact with you, or a component on the car might fail. These are all hazards that may cause you to lose control and have an accident.

Formula 1 is a high-risk sport, yet it has a very good safety record. The risks are reduced to a minimum by careful safety management, achieved through a combination of regulations, training and the use of the latest technology.

Similarly, in the workplace, the risk is the chance, high or low, that a hazard will cause someone harm. Again, these have to be carefully managed by assessing the risk and reducing the hazard's potential for harm. We will now look at some of the hazards that you might come across in engineering.

Handling and processing engineering materials can be hazardous and can present a risk to you or people around you. When you work in engineering, you need to be fully aware of these hazards and risks to health. Some of the hazards typically associated with engineering activities are listed below.

- Dust from grinding can be an irritant to the respiratory system if breathed in.
- Acids and alkalis can cause burns if they come into contact with the skin.
- Welding can produce toxic fumes, which could be breathed in.
- Misuse of tools and equipment could cause injury to personnel.

## Activity

**Find the 15 hazards in the wordsearch.**

| F | L | U | I | D | P | O | W | E | R | A | B | T | G | J | C | L | S | B | W | R |
|---|---|---|---|---|---|---|---|---|---|---|---|---|---|---|---|---|---|---|---|---|
| Q | E | T | Y | U | U | O | O | V | P | P | J | P | A | S | L | O | I | T | P | U |
| M | A | C | H | I | N | E | R | Y | F | I | O | R | Q | H | J | O | E | S | I | J |
| A | F | F | B | L | I | C | K | R | N | O | H | T | U | A | P | S | V | Y | P | C |
| N | E | L | Q | T | P | B | I | E | Q | R | S | H | F | R | E | E | N | J | O | H |
| U | T | K | E | K | C | O | N | F | I | N | E | D | S | P | A | C | E | S | Y | E |
| A | G | T | Q | U | A | N | G | K | K | Y | D | U | H | T | W | A | M | A | T | M |
| L | I | E | O | F | G | P | A | L | T | E | R | S | F | O | N | B | R | E | N | I |
| H | U | L | D | M | Z | X | T | W | A | S | T | T | N | O | O | L | E | V | A | C |
| A | L | N | W | U | E | K | H | G | N | M | T | E | I | L | A | E | N | D | O | A |
| N | O | I | S | E | R | J | E | G | Q | Y | U | I | P | S | M | S | Q | U | I | L |
| D | D | K | U | A | S | L | I | P | P | E | R | Y | S | U | R | F | A | C | E | S |
| L | G | U | Q | W | H | R | G | H | I | T | E | B | K | U | T | L | O | E | V | T |
| I | T | Y | F | S | U | R | H | L | P | G | N | K | Y | T | U | P | C | A | Z | X |
| N | E | R | E | L | E | C | T | R | I | C | I | T | Y | N | P | O | Y | K | L | Q |
| G | L | Z | H | O | K | R | I | J | B | R | L | O | S | T | O | M | Q | Y | U | R |
| X | R | O | T | A | T | I | N | G | E | Q | U | I | P | M | E | N | T | E | J | G |

## Electrical hazards

Perhaps one of the most dangerous hazards is one that we cannot see but one that we can certainly feel the effects of – electricity. It is used all around us every day and we take for granted that it is safe and available. However, we should not take for granted how dangerous it can be. If the human body comes into contact with mains or high-voltage electricity, then there is the potential for death or serious injury from electric shock.

If an electric shock causes a person to lose consciousness, then they may fall and cause further injury to themselves. The electricity flowing through the human body can also cause severe burns at the point of contact. There are a number of factors that will determine the effects of an electric shock on the human body. For example, the level of voltage and current, the duration of exposure to the shock, the person's build or size, the person's footwear and the part of the body that is affected. The safest way to approach electrical safety is to treat all electrical and electronic circuits with caution, ideally avoiding contact with them.

The mains electrical supply in a UK home is nominally 230 volts. This powers your lighting and everything else plugged into a mains power outlet, such as your TVs, games consoles, computers and refrigerators. All mains-powered electrical equipment should be considered as potentially dangerous. There is a risk of electric shock and, as such, all electrical equipment should be serviced only by trained and competent personnel.

In industrial workplaces and factories, electricity provides power for lighting, machines, tools and equipment. Unlike in your home, however, in industry it is quite common for electrical motors used to drive fixed equipment (such as fans, pumps and conveyor belts) to use a 400-volt supply. As a consequence, the risk of serious injury or death from electrocution increases if a person should receive an electric shock.

For even larger equipment used in some heavy industries, such as mining or quarries, then it is not unusual for motors to be used that require a 3,300-volt electrical supply. To provide this voltage, there will usually be an electrical substation nearby and this could have as much as 11,000 volts supplied to it. Personnel required to work on equipment involving such high voltages are specially trained and strict work-permit procedures are required.

It is not just large industrial equipment we should be concerned about when dealing with electrical safety. There are many portable hand tools that are mains powered, such as drills and grinders. On engineering sites and in workshops, portable electric tools should preferably be operated by 110 volts. As with all equipment, you should check that they are safe to use before doing so. Take any damaged equipment out of service immediately and report this to your supervisor or safety officer.

## Activity

Portable Appliance Testing (PAT) is carried out on portable electrical appliances and equipment to ensure they are safe to use. Make a list of portable electrical appliances and equipment that would be PAT tested.

## Working at height

Consider this: falls from heights are one of the main causes of death and injury in the workplace. They accounted for 46 fatal accidents and 3,350 major injuries in the year 2005/06.

Imagine yourself working from a ladder, changing a light fitting on a wall. What are the hazards and risks in this situation? Because you are working at height, there is a risk that you might fall to the ground.

**Team Talk**

Aisha: **'If falls from heights kill and injure so many people at work each year, isn't there a regulation to reduce the chances of these accidents happening?'**
Steve: **'That's right – the Work at Height Regulations 2005 were created to prevent deaths and injuries caused by falls at work.'**
Aisha: **'OK, I better have a read through those before I do the next activity.'**
Steve: **'You can view the regulations and download them from the HSE website.'**

**Activity**

Consider a task where you are required to change a light bulb in a light fitting on a wall approximately six metres (about 20 feet) above the ground. Make a list of the risks involved. What hazards might you come across? Can you suggest a safe method of completing the activity?

## Working in a confined space

A number of people in the UK are fatally injured through accidents involving confined spaces. A confined space is any space of an enclosed nature where there is a risk of death or serious injury from the effects of hazardous substances or hazardous conditions.

Many industrial processes use chemicals in the form of liquids, gases, solids, powders or granules. These are generally stored in large vessels called

**DANGER**
**CONFINED SPACE**
**HAZARDOUS ATMOSPHERE**
**CHECK OXYGEN LEVEL BEFORE AND DURING ENTRY**

**Figure 1.18** Confined space sign

tanks, vats or silos. It is possible that these products could produce dangerous gases inside the vessels. This would make the atmosphere inside them unbreathable, therefore leading to suffocation. Before any work activity is carried out in a confined space, a full and thorough safety check is made to assess the possible hazards. No work can be undertaken inside these vessels until the atmosphere inside has been checked to ensure there is sufficient oxygen. It is not only the people working in confined spaces who are killed through accidents. Rescuers who try to save them have also been suffocated because they did not take the correct safety precautions before entering the vessel.

Remember, confined spaces not only include chemical storage tanks but they could also include ductwork, sewer drains, ships' cargo holds, poorly ventilated rooms or excavated holes below ground level.

## Fluid-power safety

A fluid is something that flows. For example, water is a fluid, air is a fluid and oil is a fluid. In a fluid-power system, we take a substance that flows (usually oil or air) and direct the flow to where it is useful. Fluid-power systems are more commonly known as pneumatic and hydraulic systems. Pneumatic systems use air as the fluid and hydraulic systems use oil.

> **❝ Team Talk**
>
> Aisha: **'So fluid-power is a method of transferring energy.'**
> Steve: **'Yes, but we need to be aware of one major safety point. While the fluid is transferring its energy to produce work, it has the potential to release this energy accidentally if there is a leak.'** ❞

Fluid-power systems can use very high-pressure fluids. The pressure from your water tap at home is typically about 2 bar; a hydraulic fluid-power system can be over 200 times this pressure. There is the potential for serious injuries to occur from uncontrolled releases of high-pressure fluids. It is important that only trained and competent people maintain and operate fluid-power equipment.

# Risk assessment

M1

The Management of Health and Safety at Work Regulations 1999 set out specific requirements for health and safety. The main impact of these regulations is that an employer is required to carry out a risk assessment on certain work activities; where there are five or more employees, the employer must record the significant findings of the risk assessment.

So what is a risk assessment? This is a procedure whereby a series of steps are taken to assess any risk involved to a person carrying out an activity. Its purpose is to make sure no one gets hurt or becomes ill at work.

We each carry out risk assessments throughout our daily lives, sometimes without thinking about it. How many times have you walked across a road? Did you remember to stop, look and listen? This is a type of risk assessment.

When an employer carries out a risk assessment, they follow five basic steps.

1. Identify hazards.
2. Decide who might be harmed and how.
3. Evaluate the risks and decide on precautions.
4. Record the findings and implement any corrective actions.
5. Review the assessment and update when necessary.

It is important to recognise that there will be significant differences in the hazards associated with a typical office environment compared with a large chemical plant, for example. The risk-assessment process should not be overly complex in any case.

A guide to risk assessment is available from the HSE website (www.hse.gov.uk). As part of your studies, you will be required to carry out a risk assessment associated with a work activity, so it is advisable to obtain and read further information.

Figure 1.19 shows a typical risk-assessment form for an engineering workshop. It shows how the five steps to risk assessment are carried out and what is recorded on the form.

| Step 1 | Step 2 | Step 3 | Step 4 | Step 5 |
|---|---|---|---|---|
| Identify hazard | Decide who might be harmed and how | Evaluate the risk Decide on precautions | Record your findings | Review risk assessment Update if necessary |
| Possible slips and trips | Workshop staff and visitors may suffer injuries if they slip on uncleaned floor spillages or trip over objects and fall | Floors generally clear from objects and spillages due to good housekeeping procedures Lighting levels good Maintain good housekeeping procedures All personnel in workshop must wear safety footwear | No further action required at present | |
| Moving heavy objects/ equipment around workshop | Possibility of serious injury to workshop personnel/visitors from falling heavy objects/equipment | One technician and two operators are currently trained to move equipment safely using appropriate lifting equipment | All workshop engineering personnel to attend one-day lifting equipment course | Review due in four weeks to assess if completed |
| Machinery | Lathe, pedestal drill and milling machine operators may suffer serious injuries from unguarded machinery | All dangerous parts of machinery have guards fitted to manufacturers' standards Machinery guards inspected, tested and maintained on a regular basis New machinery undergoes pre-use checks for conformity with CE standard mark | No further action required at this stage | |

**Figure 1.19** Risk-assessment form for an engineering workshop

The following technique can be used as part of the risk-assessment procedure but is more advanced. You do not have to use it, but it is useful for determining the level of risk that arises from a hazard.

Score the likelihood of an incident or accident occurring from the hazard on a scale of 1 to 5.

1. Highly unlikely: not known to occur.
2. Remote possibility: known to occur.
3. Occasional: has happened previously.
4. Fairly frequent: some known occurrences.
5. Frequent or regular occurrences.

Next, score the severity of the effect that the hazard may have on a scale of 1 to 5.

1. Minor injury.

2. Injury would cause more than three days off work.
3. Temporary incapacity/disease.
4. Permanent incapacity.
5. Fatality.

Now, using the accompanying risk chart, plot the likelihood score against the severity score to determine the level of risk and action to be taken.

The level of risk identified from the risk assessment will decide what action should be taken.

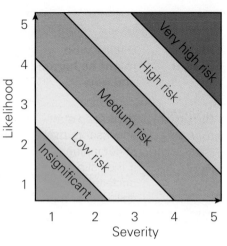

**Figure 1.20** Risk chart

| Level of risk | Action to be taken |
|---|---|
| Very high risk | Stop activity. Take immediate action. |
| High risk | Take action the same day. |
| Medium risk | Take action within one week. |
| Low risk | Monitor the situation. |
| Insignificant risk | No action required. |

**Figure 1.21** Action table

This example risk assessment is typical of one carried out for a small engineering business. It would not be suitable for all industries and engineering workplaces. A risk assessment needs to be carefully thought out and should be appropriate for the hazards and control measures required for a particular business.

## Activity

**Now that you have some background information on hazards, risks and risk-control measures, you can carry out your own risk assessment. You can use the risk-assessment form shown in Figure 1.19 or you can produce your own. Follow the five steps to risk assessment outlined previously and remember the definitions of hazards, risks and risk-control measures. Obtain a witness statement or observation record when you have sucessfully completed this activity.**

## Make the grade

This activity will help you in achieving the following grading criterion:

 **carry out a risk assessment on an engineering workplace to make recommendations on the safety of materials and equipment handling, use of personal protective equipment and the potential hazards in the area.**

## Key words

A **hazard** is something that can cause harm, illness or damage to health or property.

A **risk** is the likelihood that harm, illness or damage from a hazard will occur, and the seriousness of the consequences, e.g. how many people could be affected and how badly.

A **risk-control measure** is something put in place to eliminate or reduce exposure to a hazard.

# Warning signs

Take a relatively short car journey and you will spot many road signs. They may have different colours and shapes but they are all there to provide information. They help us to travel safely and reach our destination while staying within the law. Look at the examples of road signs in Figures 1.22–1.24. The most significant difference between them is the colours used.

Figure 1.22 means you must not exceed the 30 m.p.h. speed limit. Figure 1.23 means you must turn left. Figure 1.24 is a warning of a diversion to the normal route.

**Figure 1.22** Road sign 1

**Figure 1.23** Road sign 2

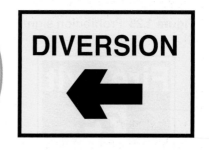
**Figure 1.24** Road sign 3

Safety signs you will come across in industry also have a colour-code system. This is to help you understand the meaning of the sign. The main types of signs are outlined below.

## Prohibition

When you see a sign with white text on a red background, it means you are prohibited from doing something. The sign in Figure 1.25 means you must not remove the guards.

## Mandatory

When you see a sign with white text on a blue background, it means you must do what the sign tells you. The sign in Figure 1.26 indicates that eye protection must be worn.

## Warning

When you see a sign with black text on a yellow background, it is giving you a warning. The sign in Figure 1.27 shows a warning of the risk of electric shock.

**Figure 1.25** Prohibition sign

**Figure 1.26** Mandatory sign

**Figure 1.27** Warning sign

**Figure 1.28** Safe condition sign

**Figure 1.29** Fire sign

## Safe condition

When you see a sign with white text on a green background, it is displaying a safe condition. The sign in Figure 1.28 indicates the safest route to take in a fire evacuation.

## Fire

Fire signs indicate the location of firefighting equipment. They have white text on a red background. The sign in Figure 1.29 indicates the location of a fire-alarm call point.

# Material and equipment handling

**P1** (Part)

In a workshop, warehouse or storage area, it is usually necessary at some stage to move equipment and materials around. It does not matter whether the load to be moved is a box, a machine, a person or an animal; general principles of safe manual handling apply. The Manual Handling Operations Regulations 1992 (amended in 2002) apply to a wide range of manual handling techniques, such as lifting, lowering, pulling, pushing or carrying. They are in place to help prevent injuries from manual handling.

### Activity

Review safe manual handling procedures using information available from www.hse.gov.uk/pubns/indg143.pdf

Ideally, an employer should try to avoid the need for hazardous manual handling if at all possible. If manual handling cannot be avoided, the risk of injury has to be assessed and reduced as far as is reasonably practicable. As an employee, you need to follow any safe systems of work set out for your safety and use any equipment provided for the task.

Let us go through a typical procedure for planning and carrying out the movement of materials or equipment into a workshop. We need to consider how this can be done as safely and effectively as possible. Good preparation is the key.

The initial planning for the task requires the involvement of all the personnel involved with the job. Your supervisor or Health and Safety Officer should be aware of the task being carried out. If required, ensure a permit to work is in place. People in the work area should be informed of the movement of materials near them.

Next, check your personal protective equipment is suitable for the task. Local area safety measures will apply and signs will also indicate what specific PPE is required. Typically for a task like this, you will need a hard hat, safety shoes, gloves and overalls. Decide on the best route the equipment or material should take, giving consideration to people working in the area. The movement of materials is made more difficult by obstructions that cannot be moved.

If the materials or equipment are particularly bulky or heavy, then it will be necessary to use a mechanical lifting aid. These assist a person in lifting and transporting objects that otherwise could not be moved by just one person. There are many different types of lifting equipment available and there is no such thing as a completely safe manual handling operation. You also need to be aware that using lifting or handling aids might introduce new hazards.

## Handling materials and equipment safely – basic lifting and carrying technique

Before moving an object, you need to make a basic assessment of whether you can manage with or without lifting aids. Try tipping the load slightly to determine whether you are capable of moving it. Another method is to estimate the weight based on the object's volume. The maximum safe-lifting capability depends on whether you are male or female. Guidelines are shown in Figure 1.30. The weights are reduced if the lifting operation is done

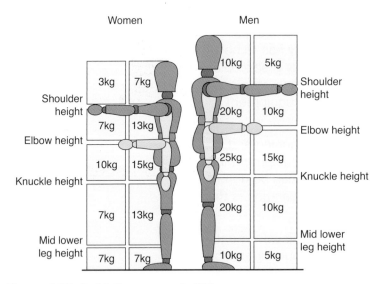

**Figure 1.30** Guidelines on safe lifting

frequently or if it is necessary to twist when the load is lifted.

Basic safe manual lifting techniques are shown in Figure 1.31.

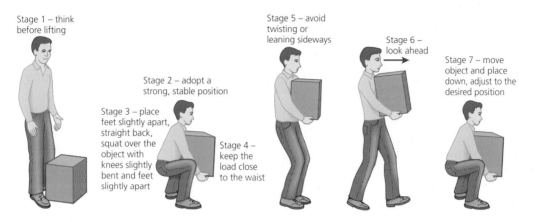

Stage 1 – think before lifting

Stage 2 – adopt a strong, stable position

Stage 3 – place feet slightly apart, straight back, squat over the object with knees slightly bent and feet slightly apart

Stage 4 – keep the load close to the waist

Stage 5 – avoid twisting or leaning sideways

Stage 6 – look ahead

Stage 7 – move object and place down, adjust to the desired position

**Figure 1.31** Safe manual lifting techniques

## Handling materials and equipment safely – manual lifting aids

If the load cannot be moved safely using basic manual handling methods, then manual lifting aids can be used. Some examples of these are shown in Figure 1.32.

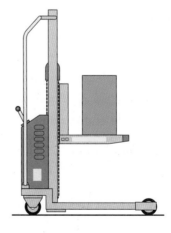

**Figure 1.32** Manual lifting aids

# Handling materials and equipment safely – mechanical lifting aids

Some engineering facilities, such as large process plants, have structures that are tall and difficult to access. It may sometimes be necessary to lift and lower equipment to and from these structures. For these activities, a full risk assessment should be carried out and specialised lifting equipment would be required. Fibre slings and wire ropes can be used in conjunction with lifting gear, such as cranes, chain blocks and pulleys. Lifting equipment must be in good condition before use. It is a requirement under the Lifting Operations and Lifting Equipment Regulations (LOLER) 1998 that lifting equipment should be inspected for defects and tested for safe working load on a regular basis.

These basic steps for safely moving equipment and materials should be followed whether the work area is a workshop, warehouse, building site or large industrial premises.

**Figure 1.33** Mechanical lifting aids

## Make the grade

This activity will help you in achieving the following grading criterion:

**P1** (part) handle materials and equipment in an engineering workplace in a safe and approved manner.

## Activity

1. Explain the correct methods required for handling and moving materials and equipment for the activities below. (Your tutor or supervisor may select a different activity for you.)

   - Collect and transport cardboard boxes full of industrial cleaning cloths from the store to the workshop area.
   - Secure a workpiece onto a pedestal drill table using a vice or clamps and then safely complete a drilling activity.
   - Change a lathe chuck safely (e.g. from a three-jaw type to a four-jaw or split collet type).
   - Take a long and heavy piece of material to a bandsaw ready for cutting into smaller lengths.
   - Transport several small but relatively heavy items (toolboxes or cable drums could be used for this) through the workshop area.

2. Using one of the examples from the list above (or one selected for you), carry out the activity, demonstrating you can handle materials and equipment in a safe and approved manner. Obtain a witness statement or observation record to show that you have completed this activity.

# Health and safety policies

Any company that employs five or more people must, by law (Health and Safety at Work Act 1974), have a written health and safety policy. This is a document that records the responsibilities of the organisation and the arrangements it makes to ensure the health and safety of the employees.

The health and safety policy must explain how an organisation manages health and safety by defining who does what, when they do it and how they do it.

A general statement from a typical health and safety policy document is shown in Figure 1.34.

---

**General Health and Safety Policy Statement of:**
**A Company Limited**

---

- To provide adequate control of the health and safety risks arising from our work activities.

- To consult with our employees on matters concerning health and safety.

- To provide and maintain safe plant and equipment.

- To ensure substances are handled and used safely.

- To provide information, instruction and supervision for employees.

- To ensure all employees are competent to carry out their duties and to provide adequate training.

- To prevent accidents and instances of work-related ill health.

- To maintain safe and healthy working conditions.

- To review and revise the health and safety policy at regular intervals.

| Signed: | Employer: |
|---------|-----------|
| Date | Review date: |

**Figure 1.34** A general statement from a typical health and safety policy document

## Make the grade

The next activity will help you in achieving the following grading criterion:

 prepare a safety policy for an engineering work area including references to relevant legislation.

## Activity

Prepare a complete health and safety policy document relating to an engineering work area. Use the general health and safety policy statement example in Figure 1.34 to get you started as well as relevant regulations. The health and safety policy should set out the roles and responsibilities for specific health and safety areas. You can use the following sections to ensure you have included all the relevant information.

**Roles and responsibilities**

In this section, you need to include the roles and responsibilities given to specific personnel in the company regarding health and safety matters. For example, you should name the person who would have overall or complete responsibility for an engineering workshop and also identify those who would have responsibility for a specific area.

**Risk-assessment arrangements**

This section essentially covers who has responsibility for any health and safety risks arising from work activities. This section needs to include who will carry out any risk assessments, the person to whom the results of the risk assessments are reported and who approves, implements and reviews any corrective actions taken.

**Consultation with the workforce**

This section is used to show that the employees have been consulted with regard to matters on health, safety and welfare. In this section, you can include the name or names of employee representatives or trade union-appointed representatives.

**Safe plant and equipment**

In this section, you should name the personnel who have responsibility for maintaining plant and equipment so that they meet health and safety standards. This includes existing plant and equipment and also any new or used items.

**Safe handling and control of substances**

This section is related to all the responsibilities under the Control of Substances Hazardous to Health Regulations (COSHH) 2002. Identify and list who is responsible for identifying hazardous substances, carrying out the COSHH assessments, implementing any corrective actions and relaying information to employees.

**Information, instruction and supervision**

Information about where employees can go for information and advice on health and safety should be contained in this section – for example, the location of posters or leaflets on health and safety law and who is responsible for the supervision of young workers or trainees.

### General health and safety and job-specific training

In this section, you should include the names of those people responsible for providing basic safety training, such as safety inductions (these would normally be for new employees and cover such topics as first aid and fire safety) or training for specialised jobs.

### Accidents, first aid and ill health due to work

Certain work requires employees to receive special health surveillance. If this is the case, then it should be included in this section. You should also include information on: who the first aider is, the location of the first-aid facilities and who is responsible for recording and reporting accidents, diseases and dangerous incidents.

### Emergency procedures

This section should be used to record the relevant emergency procedures for an organisation. Items such as checking and maintaining fire extinguishers, fire-alarm testing, emergency evacuation and escape-route monitoring should be included here.

# Learning Outcome 2. Be able to work efficiently and effectively in engineering

## Preparing for and planning engineering activities

There is a saying, 'To fail to prepare is to prepare to fail.' This is certainly relevant where engineering activities are concerned. Before any engineering task is carried out, you should ensure that you are fully prepared.

This means knowing exactly what it is you have to do and also what tools, equipment and materials you are going to need. The tools required to complete engineering activities are not always just the hammers, screwdrivers and other items from your toolbox. Documentation such as plans, specifications, technical manuals, engineering

**Figure 1.35** Strategy and planning

drawings and reports are just as important and sometimes required.

Even a typical engineering activity, such as drilling a hole in a piece of metal using a pedestal drill, requires some forethought and preparation. So consider the preparation and planning required to overhaul a large oil refinery or food factory. We will now look at what you should be doing when preparing for a work activity.

First of all, you should make sure you are certain about what it is you are being asked to do. In some industries, a written job card or instruction sheet tells you exactly what the job will involve. You should only carry out work activities that you have been authorised to complete. This means you have been trained or instructed in how to complete them safely.

Check the work area is safe, clean and tidy and ensure you have the correct PPE for the task. You should then gather all the tools, materials and equipment required and examine them for faults or damage before starting the activity. Once you have completed the engineering work activity, you should ensure all tools, materials and equipment are returned to the proper storage location and the work area is left as you found it: safe, clean and tidy. You may have further documentation to complete, such as signing off a permit or completing a logbook or report. You may even want to suggest ways to improve the activity for the next time.

## Make the grade

This activity will help you in achieving the following grading criterion:

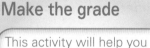 **P5**    prepare for and carry out an engineering work activity.

## Activity

1. **Make a list of the tools, materials and equipment you would need to prepare in order to complete a typical engineering workshop activity. Some examples of engineering work activities are given below. However, you may choose your own or your supervisor may suggest others for you.**
   - **Drilling material safely using a workshop pedestal.**
   - **Turning material using a centre lathe.**
   - **Wiring electrical and electronic circuits.**
   - **Building a mechanical assembly.**
2. **Carry out an engineering work activity using one of the examples above (or one selected for you). Obtain an observation record from your supervisor or tutor indicating you have safely completed the engineering work activity.**

## Activity

Suggest ways of improving the work activity you carried out for the P5 activity on page 36.

### Make the grade

This activity will help you in achieving the following grading criterion:

 **M3** identify how a work activity could be improved.

# The work environment

Work in the engineering sector encompasses a wide range of occupations. Your work environment and working conditions could vary according to your role. You could work in an office, in a laboratory, on a factory floor, in a workshop, on a site outdoors or any combination of these.

To keep your work environment safe, it is important that you keep your work area clean and tidy. Have to hand only the tools and equipment that you will need. You may often see or hear the phrase 'Tidy up as you go.' This means you should continuously maintain the tidiness of the work area and not just clean up at the end of each job. Any unnecessary clutter will make the work area untidy and present tripping hazards. Unseen spillages could cause a slipping hazard if not dealt with straight away and also present a potential fire or environmental pollution risk. Once you have finished the task, you should clean, inspect and return any tools or equipment to their storage point.

As an engineer, you could be involved with manufacturing and production, or operations and maintenance. This means you could be making sure equipment is kept running efficiently and effectively. Regardless of the area you work in, many of the activities associated with engineering can be potentially hazardous. If you are unsure of how to carry out an activity safely, there are people you can ask for further information and clarification. There will also be other sources of information to guide you, such as safety posters and warning signs.

## Activity

Make a list of the people you could ask for further information and guidance if you were unsure how to complete an engineering activity safely.

Let us now take a closer look at some potentially hazardous activities in engineering sectors.

## Manufacturing and production

In the engineering sector, you may come across activities where material removal is taking place, using machines such as grinders, milling machines, lathes or drills. Activities such as these produce hazards from hot and sharp fragments (sometimes called swarf) being ejected from the cutting tool. Guards are used on these machines to prevent personnel from coming into contact with swarf, rotating parts, cutting tools, bar stock or drive belts. You must never remove or interfere with a guard on a machine.

We mentioned previously that it is not only the user of the machine or equipment who has a responsibility under the HASAWA 1974. The designer of the machine has a responsibility to ensure it is designed to be safe when properly used. Some guarding systems on machines have electrical safety interlocks. These prevent the machine from operating until the guard has been put in place. In spite of this, many accidents have occurred because the interlock has been defeated. In these cases, the blame transfers from the employer to the employee when the latter improperly removes the guard.

Irrespective of the quality of the design of safeguards, long-term accident prevention depends on:

- safe systems of work, which allow the supervised removal of safeguards only when it is safe to do so, or when all reasonable steps have been taken to minimise the risk;
- regular maintenance and inspection of safeguards by competent personnel.

Injuries caused by machine components generally fall into the following five groups:

- traps;

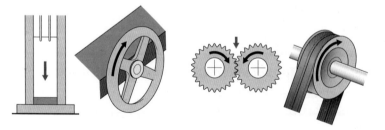

**Figure 1.36** Traps

- impact;

Impact

**Figure 1.37** Impact

- contact;

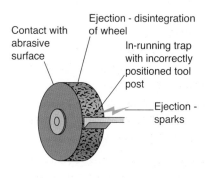

Contact danger

**Figure 1.38** Contact

- entanglement;

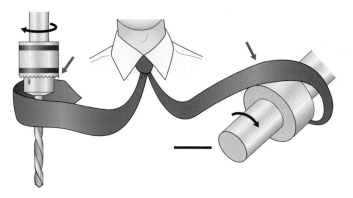

**Figure 1.39** Entanglement

- ejection.

Ejection - disintegration of wheel

Contact with abrasive surface

In-running trap with incorrectly positioned tool post

Ejection - sparks

Hazards presented by an abrasive wheel

**Figure 1.40** Ejection

Material joining, forming or casting operations present hazards such as noise, fumes and danger of burns from hot surfaces. Some engineering processes require the storage and use of hazardous chemicals such as acids, alkalis or solvents. To prevent contact with these chemicals, the correct personal protective equipment (PPE) must be worn.

In all cases, you should receive training on how to complete the activity safely. There should also be warning signs in the workplace, indicating any hazards and the appropriate safety measures to take. Access to areas with hazardous processes should be controlled and restricted to relevant personnel.

## Operations and maintenance

There are certain areas of industry that operate continuous production processes. For example, chemical plants, oil refineries, oil rigs and power stations run almost constantly. The only time they shut down is for scheduled maintenance, although occasionally equipment can break down unexpectedly.

Large factories and commercial buildings require heating, ventilation and air conditioning (usually abbreviated to HVAC) to operate almost constantly to maintain a comfortable environment, and hospital operating theatres and micro-electronics factories require the building's internal environment to be carefully controlled.

The typical systems an engineer may be expected to operate and maintain are:

- low- and high-voltage electrical supplies;
- pressurised pipework and vessels;
- rotating machinery;
- heating, ventilation and air conditioning.

The engineering systems associated with these areas of industry are not always hazardous in themselves. However, they could be when used in hazardous environments. Think in terms of areas where flammable liquids, gases or corrosive/radioactive substances could be present. Due to these hazards, safe systems of work must be in place. Before any system or equipment is made available to the engineer, rigorous checks must be carried out to ensure it is safe.

One such safety system is known as a permit-to-work (PTW) system. Before an engineer is allowed to commence work on a piece of equipment or plant system, a PTW must be completed. This document provides details of how the equipment or plant system has been made safe, ready for the work to commence.

Using some of the previously listed examples, the PTW would be used to record that electrical power supplies were isolated to prevent electrocution, that pressurised systems were vented to release pressure and/or flammable/noxious gases and that rotating machinery was stopped and electrically isolated.

## Working relationships

There are not many careers where you work entirely on your own. Even a long-distance lorry driver or travelling sales representative will need to communicate with their manager or customers. We all work as part of a team and how well these teams work can determine the overall effectiveness of a business.

We all know that when teams work well together, fantastic results can be achieved. You only need to look at some of the great sporting achievements in recent history to realise how effective good teams can be.

Good teams require the team members to work well together. For this to happen, team members need to have a good working relationship. It is generally accepted that in a happy workforce there are good working relationships in place. In a workplace environment, this also leads to high morale and greater productivity.

**Figure 1.41** Good working relationships are important

To build a good working relationship requires cooperation and trust. This is not always easy. Each individual is unique and, as such, has differing emotions, needs and personal objectives. A good way of building positive working relationships with your colleagues (or even customers) is to establish and maintain good communication with them. It is also important to listen to what they have to say. Everyone is entitled to their opinions even if you disagree with them.

In an engineering environment, good working relationships are essential. A poor working relationship in any workplace could have many negative repercussions on productivity, efficiency and health and safety.

Strong working relationships come from respecting your team members and colleagues. You should aim to work towards a common goal. Most people can identify problems with systems or in the way work activities are carried out. To earn respect from your peers and your boss, you should also suggest solutions to these problems. Good communication also contributes to strong working relationships. You should try to be positive and share credit for ideas, accomplishments and contributions.

We will now consider what sorts of things could have a negative effect on a working relationship. Perhaps the most common cause of a breakdown in a working relationship is conflict between individuals. There are many different reasons why conflict occurs; sometimes it can be because of seemingly trivial issues.

In a team, when we work alongside each other in close proximity for any length of time, conflict is inevitable.

Disagreements can occur over minor as well as significant issues. It is how we react and deal with these disagreements that is important.

The best way of dealing with conflict is to try not to overreact. Make sure you are clear in your mind what you are objecting to and that your feelings are in proportion to the disagreement. Other people are then more likely to consider your viewpoint and you will be able to listen to their viewpoints more accurately. Ideally, find a solution that meets the needs of all those involved. Remember, this is not always possible, so you may need to accept a compromise.

Try to imagine how you would feel in the other person's place. It may be that, even though you do not agree with them, it will help you to understand why they are feeling the way they do.

If the disagreement cannot be resolved amicably, then a third party should be informed. It might help to discuss it with your boss or with a third party who can be trusted and who is willing to become involved in settling the disagreement.

## Activity

Write down how you would feel about the following issues and how you think they should be dealt with effectively.

1. Someone borrowed your tools without asking and returned them dirty and damaged.
2. You were told to clean up a mess in the work area that your colleague left previously.
3. Your workmate always does a job in a certain way and you think you could do it better and quicker using your own method.
4. One of your team members always receives praise from the supervisor or manager, yet you never seem to get any praise.
5. A customer makes a request for information that you do not have.
6. You work in a team and you all do the same job. However, each team member gets paid a different salary and you believe you are earning the least.
7. One of your team members always arrives late for work and often takes sick leave.
8. One of your colleagues always pokes fun at you and other colleagues.

### Make the grade

This activity will help you in achieving the following grading criterion:

 **P6** maintain good working relationships with colleagues and other relevant people when carrying out an engineering work activity.

## Activity

Identify and write down the strengths and weaknesses of one of your working relationships. Describe what you could do to improve this working relationship.

### Make the grade

This activity will help you in achieving the following grading criterion:

**D2** identify strengths and areas for improvement in a working relationship.

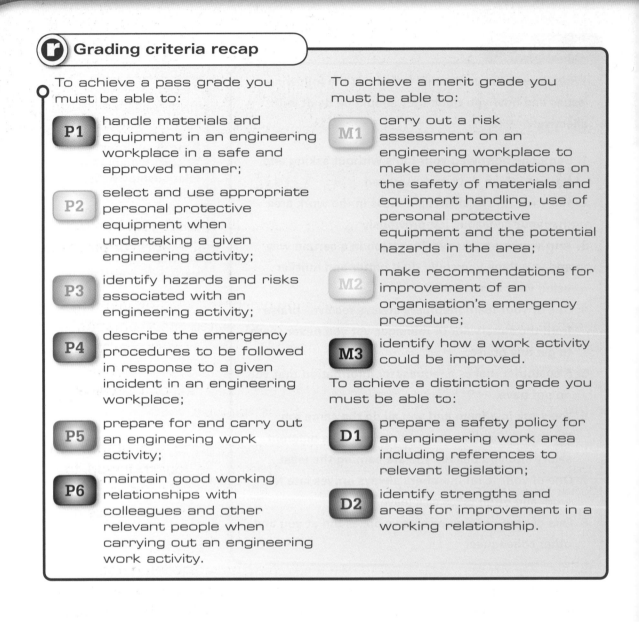

**r** **Grading criteria recap**

To achieve a pass grade you must be able to:

**P1** handle materials and equipment in an engineering workplace in a safe and approved manner;

**P2** select and use appropriate personal protective equipment when undertaking a given engineering activity;

**P3** identify hazards and risks associated with an engineering activity;

**P4** describe the emergency procedures to be followed in response to a given incident in an engineering workplace;

**P5** prepare for and carry out an engineering work activity;

**P6** maintain good working relationships with colleagues and other relevant people when carrying out an engineering work activity.

To achieve a merit grade you must be able to:

**M1** carry out a risk assessment on an engineering workplace to make recommendations on the safety of materials and equipment handling, use of personal protective equipment and the potential hazards in the area;

**M2** make recommendations for improvement of an organisation's emergency procedure;

**M3** identify how a work activity could be improved.

To achieve a distinction grade you must be able to:

**D1** prepare a safety policy for an engineering work area including references to relevant legislation;

**D2** identify strengths and areas for improvement in a working relationship.

# Unit 2
## Interpreting and using engineering information

# Introduction to the unit

This unit will help you gain the knowledge and skills that are required to use engineering information when carrying out a manufacturing or process operation.

The unit is split to cover both of the learning outcomes. The first learning outcome looks at 'how to interpret drawings and related documentation', examining how to extract information from various sources so that tasks can be carried out. The first part of the unit explores the different types of engineering drawings in detail to gain an understanding of the typical information they convey and the related documentation.

The second learning outcome gives you the opportunity to 'use information from drawings and related documentation'. You can use your own work to identify the information required so that it can be carried out and checked from given engineering information. You will also learn the importance of considering the care, control and security of information.

Engineering drawings are used extensively in the manufacturing and process industries. Figure 2.1 shows, at a glance, the different types of engineering drawings that will be covered in this unit. These drawings are shown together to give you a visual appreciation of the different types of engineering drawings that you are likely to encounter and that you will need to understand to work in an engineering environment.

## Learning Outcomes

By the end of this unit you should:

- know how to interpret drawings and related documentation;
- be able to use information from drawings and related documentation.

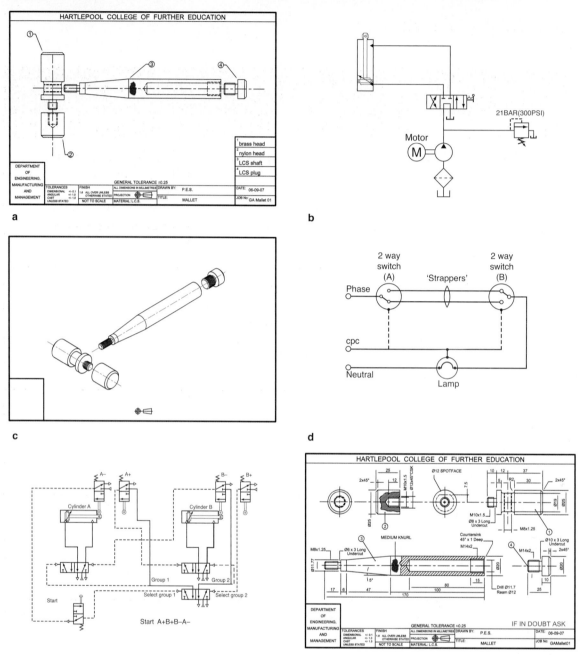

**Figure 2.1** Different types of engineering drawings: a) assembly drawing; b) hydraulic drawing; c) exploded drawing; d) electrical drawing; e) pneumatic drawing; f) detailed drawing

# Grading criteria

| To achieve a pass grade you must be able to: | To achieve a merit grade you must be able to: | To achieve a distinction grade you must be able to: |
|---|---|---|
| **P1** extract information from engineering drawings and related documentation to enable a given task to be carried out | **M1** identify gaps or deficiencies in the information obtained that need to be resolved to enable a given task to be carried out | **D1** justify valid solutions to meet identified gaps or deficiencies with the information obtained |
| **P2** select and use other information sources to support and check information provided | **M2** identify improvements in the care and control procedures used for drawings and related documentation | |
| **P3** identify and obtain relevant drawings and related documentation to carry out and check own work output | | |
| **P4** complete all necessary production documentation related to own work output | | |
| **P5** describe the care and control procedures for the drawings and related documentation used when carrying out and checking own work output | | |

# Learning Outcome 1. Know how to interpret drawings and related documentation

## Extracting information from engineering drawings and related documentation

P1

If you want to extract information from an engineering drawing, then you need to understand the basics. A typical formal engineering drawing should have a border and a title block (usually displayed along the bottom or the right-hand side of the drawing). Complex engineering drawings usually have letters (vertical or y axis) and numbers (horizontal or x axis) set out as a grid so that specific parts can be quickly identified, similar to a map. The fundamental information that is included in the title block (shown in Figure 2.2) is as follows:

1. The title or name of the drawing.
2. The name (and sometimes signature) of the person who created the drawing.
3. The drawing number (this can be issue number, job number or number of the revised version of the drawing).
4. The date that the drawing was completed.
5. The type of material(s) for the drawn component.
6. The scale of the drawing (can sometimes indicate 'not to scale').
7. The unit of measurement (mm or inches).

Note – different companies or establishments have their own style for a title block, so they can differ from the example shown in Figure 2.2, but they should include the seven pieces of information highlighted above.

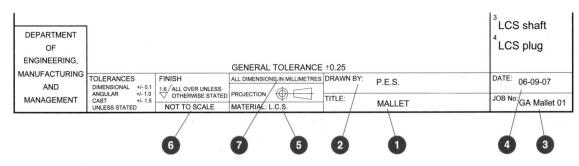

**Figure 2.2** Title block

## Activity

Use a piece of A4 paper to practise laying out an engineering drawing (this can be done roughly at this stage, so you only need a pencil and ruler). Include the following:

- your name;
- drawing number (No. 1);
- drawing title or name;
- the date;
- the scale of the drawing (in this case, 2:1);
- dimensions in millimetres.

Now measure your mobile phone in millimetres (the length and width of the phone and the screen), then double the sizes because the scale is 2:1. (If it were 3:1, it would be three times the size; if it were 1:2, it would be reduced to half the size.) Once you have worked out the sizes, draw the phone in the centre of the page using a pencil and ruler. Complete as much of the title block as you can from the information you have. You could also include the finer detail of the phone, such as the buttons and logo.

Once you have completed this activity, you will have an understanding of the basic fundamentals for interpreting the title block, which is the first item you should look at on an engineering drawing.

There are various types of engineering drawings, which look similar because they tend to follow certain standards. BS8888 is the British Standard that is used to offer consistency in engineering drawings and with all related technical product documentation. The standard includes lettering, units, tolerances, flow diagrams and product specifications.

## Dimensional detail

You will be expected to use metric units of measurement (millimetres and metres) when producing an engineering drawing, but a lot of drawings used in industry have imperial units of measurement (inches and feet). This is because the drawing is either quite old or from a different country. A comparison of metric and imperial units is shown in Table 2.1; inches are in decimal format and are also expressed as a fraction to make the comparison easier to understand.

| Inches (") | Millimetres (mm) | Fraction (") |
|---|---|---|
| 0.125" | 3.175 mm | ⅛" |
| 0.25" | 6.35 mm | ¼" |
| 0.375" | 9.525 mm | ⅜" |
| 0.5" | 12.70 mm | ½" |
| 0.625" | 15.875 mm | ⅝" |
| 0.75" | 19.05 mm | ¾" |
| 0.875" | 22.225 mm | ⅞" |
| 1" | 25.40 mm | 1.0" |

**Table 2.1** Comparison of metric and imperial units

# Dimensional detail and manufacturing/production detail

Dimensional detail and manufacturing/production detail can be taken from an engineering drawing, usually from the title block. Figure 2.3 highlights some of the information that is used by the manufacturer to ensure that the product meets certain standards; this particular example is taken from Figure 2.2. The tolerance, finish and projection will now be discussed in more detail.

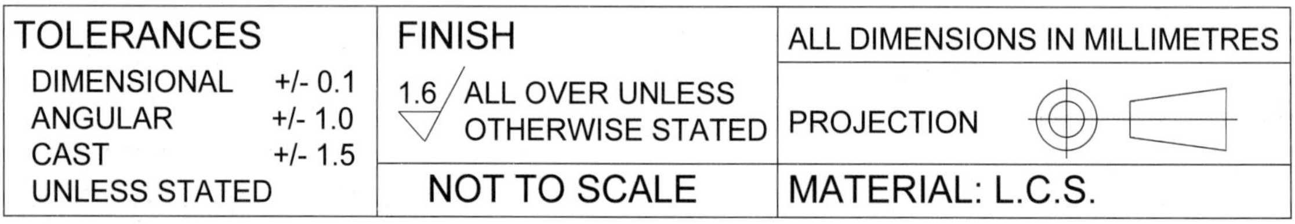

| TOLERANCES | FINISH | ALL DIMENSIONS IN MILLIMETRES |
|---|---|---|
| DIMENSIONAL    +/- 0.1<br>ANGULAR    +/- 1.0<br>CAST    +/- 1.5<br>UNLESS STATED | 1.6 / ALL OVER UNLESS<br>▽   OTHERWISE STATED | PROJECTION ⊕ ⊲ |
| | NOT TO SCALE | MATERIAL: L.C.S. |

**Figure 2.3** Title block detail

## Tolerance

It is very difficult to manufacture a product to the exact sizes on an engineering drawing, particularly if you are working with hundredths of a millimetre. Engineers overcome this by adding a tolerance to the sizes on the drawing; this means that a small amount of allowance above and below the dimensions is allowed. For example, a length of 100 mm steel bar may have a tolerance of +/− 0.5 mm. If you cut the bar and it falls between 99.5 mm and 100.5mm, then it is said to be within tolerance and is acceptable. If it were out of this range, then it would not meet the standard. Figure 2.3 has three different tolerances, which cover dimensional sizes, angles and cast components.

Linear and geometric tolerances can also be indicated on the drawing instead of the title block. A set of symbols, taken from BS ISO 1101:1983, is used to identify the type of tolerance; eight symbols that are commonly used on engineering drawings are shown in Table 2.2.

| Tolerance symbol | Meaning | Tolerance symbol | Meaning |
|---|---|---|---|
| ▱ | Flatness | ◎ | Coaxiality |
| ○ | Circularity | ⊕ | Position |
| — | Straightness | ═ | Symmetry |
| ⌭ | Cylindricity | // | Parallelism |

**Table 2.2** Tolerance symbols

The symbols are then entered into what is known as a tolerance frame, which is divided into two or more boxes. An example of this is shown in Figure 2.4, with the first box indicating the tolerance feature with the relevant symbol. The second box has the +/− tolerance value (this may be preceded with a diameter symbol ∅ if the tolerance is circular). The third box indicates the datum to which the tolerance refers.

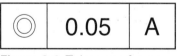

**Figure 2.4** Tolerance frame

## Surface finish and surface texture

Figure 2.3 (title block detail) shows a symbol and number for the surface finish of a component (usually referred to as 'finish'). The surface texture of a material is how rough or smooth the surface is and this can be measured using devices that record the Ra value (roughness average or mean roughness). The Ra value is discussed in more detail in Unit 14 (Selecting and using secondary machining techniques to remove material). The symbol that machinists need to interpret gives a clear indication of how the surface finish should appear (Figure 2.5).

The surface finish symbol itself also has a meaning and can be changed to indicate machining requirements, as shown in Figure 2.6.

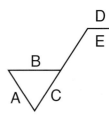

**Figure 2.5** Surface finish interpretation: a) machining allowance; b) surface finish; c) direction of lay; d) surface finish method (e.g. ground, turned, etc.); e) surface texture sampling length

## Activity

Identify the symbol shown in Figure 2.7 and explain what the five pieces of information are surrounding it.

**Figure 2.7**

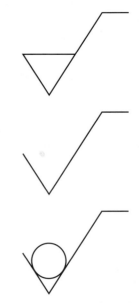

**Figure 2.6** Types of surface finish symbols: a) material removed by machining; b) basic symbol; c) no removal of material by machining

## Projection

Projection symbols indicate the projection of the engineering drawing (how the component/drawing is viewed by the reader). The two types are first-angle and third-angle projection (Figure 2.8). This is explained in more detail in Unit 10 (Using computer-aided drawing techniques in engineering).

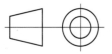

First angle projection            Third angle projection

**Figure 2.8** First-angle and third-angle projection symbols

## Other symbols and abbreviations

There are other types of symbols used in different areas of engineering and some of them are detailed in other units in this book.

Electronic symbols are used to represent electronic components on circuit diagrams. These are discussed in Unit 19 (Electronic circuit construction).

Symbols and abbreviations that are generally used on engineering drawings to identify types of materials are discussed in Unit 8 (Selecting engineering materials). There is also an extensive list taken from BS8888 on the following website: http://www.roymech.co.uk/Useful_Tables/Drawing/ABREV.html

Reference charts for limits and fits are used by machinists so that the finished product can be either tight or loose. There are three main types of fit in engineering.

- Interference fit – this is where two components are forced together as the hole would be smaller than the component entering it. A good example of this is the shaft from a hammer being forced into the head.

- Transition fit – this type of fit is used where two components neatly fit together and allow a very slight amount of movement. An example of this is a slideway on a machine tool. The two components are not forced together but must be properly aligned because it is a very close fit.

- Clearance fit – this is when two components fit together with a degree of movement between them. A golf pin in a golf hole (cup) has a degree of clearance so that it can be easily inserted and removed.

A typical reference chart for limits and fits can be used (Table 2.3), with symbols representing the hole and shaft.

| ISO description | Hole tolerance | Shaft tolerance |
| --- | --- | --- |
| Loose fit | H11 | c11 |
| Sliding fit | H7 | g6 |
| Location clearance | H7 | h6 |
| Slight interference | H7 | k6 |
| Transition | H7 | n6 |
| Force fit | H7 | u6 |

**Table 2.3** Reference chart for limits and fits

When a specific size of thread is needed in a component, machinists will often use a drill and tapping table to determine the size of the tap pilot hole. Table 2.4 shows the size of hole (tapping pilot drill diameter) that needs to be drilled for a particular thread size (nominal size of threads and pitch). This is of vital importance because sometimes the thread size is stated on an engineering drawing but the drill size is not.

| Metric (ISO) coarse | | |
| --- | --- | --- |
| Nominal diameter of threads (mm) | Pitch (mm) | Tapping drill diameter (mm) |
| M2.0 | 0.40 | 1.60 |
| M2.2 | 0.45 | 1.75 |
| M2.5 | 0.45 | 2.05 |
| M3.0 | 0.50 | 2.50 |
| M3.5 | 0.60 | 2.90 |
| M4.0 | 0.70 | 3.30 |
| M4.5 | 0.75 | 3.70 |
| M5.0 | 0.80 | 4.20 |
| M6.0 | 1.00 | 5.00 |
| M7.0 | 1.00 | 6.00 |
| M8.0 | 1.25 | 6.80 |
| M9.0 | 1.25 | 7.80 |
| M10.0 | 1.50 | 8.50 |

**Table 2.4** Drill and tapping table

The drill and tapping table (Table 2.4) represents only a small amount of information. The thread size could go up to 60 mm, could be metric fine, could be in imperial measurement, or it could be a pipe thread or another type of thread. An example of a range of drill tapping sizes can be found at the following website: http://www.goliath.com.au/Tapping%20Drill.htm.

## Welding drawings and symbols

There is a vast amount of symbols used in welding engineering drawings and they can be very complex. Some of the basic symbols will now be examined. The British Standard used to represent symbols on a welding drawing is BSEN22553 –1995.

If two metal components need to be welded together, a simple engineering drawing would indicate with the point of an arrow where the weld should be made, as in Figure 2.9 (weld shown in red).

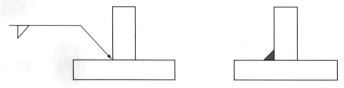

**Figure 2.9** Weld symbol 1

The reference line that is pointing to the weld area includes a symbol that means a fillet weld should be applied.

If the fillet symbol were on top of the line, the weld should be applied on the opposite side, as in Figure 2.10.

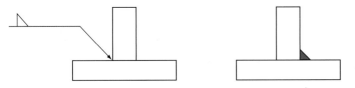

**Figure 2.10** Weld symbol 2

The reference line sometimes has a tail, which represents other information, such as a welding process or position, specification information or a non-destructive testing method. Figure 2.11 is an example of a reference line with a tail. The number 131 means that the type of welding to be used should be MIG welding.

**Figure 2.11** Weld reference line

Other basic weld symbols that can be used on drawings are shown in Table 2.5. The basic symbols represent the specific types of weld.

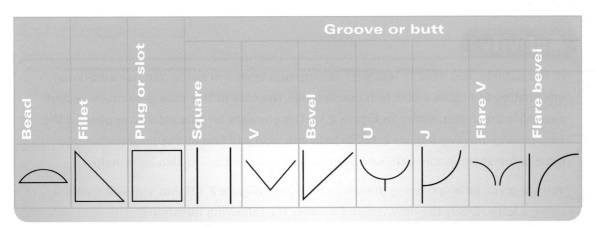

**Table 2.5** Basic weld symbols

Table 2.6 shows some of the processes that are identified by a number on a welding drawing.

| Number | Process |
|--------|---------|
| 11 | Metal arc welding without gas protection |
| 111 | MMA welding with covered electrode |
| 114 | Flux cored metal arc welding |
| 121 | Submerged arc welding, with wire electrode |
| 131 | MIG welding |
| 135 | MAGS welding |
| 141 | TIG welding |
| 15 | Plasma arc welding |
| 21 | Spot welding |
| 22 | Seam welding |
| 221 | Lap seam welding |
| 23 | Projection welding |
| 25 | Resistance butt – welding |
| 31 | Oxy-fuel gas welding |
| 311 | Oxy-acetylene welding |

**Table 2.6** Identifying numbers for different welding processes

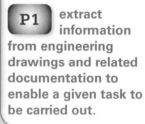

**Make the grade**

This activity will help you in achieving the following grading criterion:

**P1** extract information from engineering drawings and related documentation to enable a given task to be carried out.

## Activity

In this activity, you need to interpret information from a drawing and use additional information to enable a task to be carried out. The task in this case is to machine and assemble the mallet shown in Figure 2.13. The answers for this activity appear on the engineering drawing (Figure 2.12). To help you, you will find all the information you need somewhere in this unit; additional documentation could also be of help.

You are given an engineering drawing of a mallet (Figure 2.12) that you will machine using a lathe. Study the drawing and answer the following questions.

1. What should the overall surface finish of the mallet be?
2. Name four machining instructions indicated on the drawing. In what units are the dimensions measured?
3. What is the symbol used to denote the diameters on the drawing?
4. If the raw material were a casting, what would the tolerance allowance be?
5. Name the three different-sized threads on the drawing (including the pitch).
6. If only imperial bar stock were available, what sized bar would be required for the mallet shaft (part number 3)?

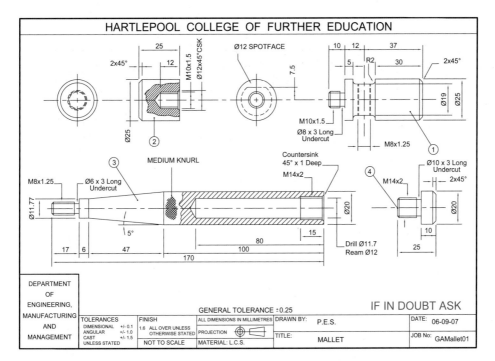

**Figure 2.12** Engineering drawing of a mallet

# Types of engineering drawings

## General assembly drawings

General assembly drawings and sub-assembly drawings are used to show how a product is assembled. They can be fully assembled or they can be laid out to show how a product fits together. The different parts of the product are sometimes listed in a separate table and are represented on the drawing with a number.

Figure 2.13 shows an assembly drawing of a mallet that is made with four different parts and three different types of material (brass, nylon and low-carbon steel). It is not possible to identify the different materials from the drawing, so they will be listed on the drawing, in a table or on additional documentation.

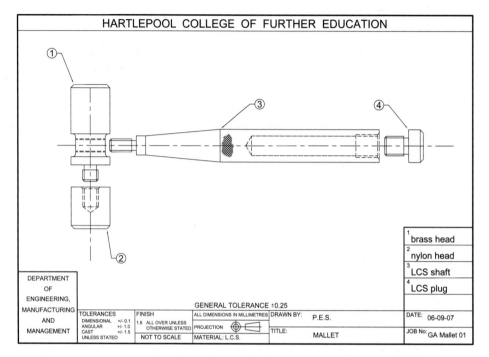

**Figure 2.13** General assembly drawing of a mallet

## Detailed drawings

Detailed drawings should include enough detail so that the product can be manufactured. Figure 2.14 shows a detailed drawing of the mallet from Figure 2.13. Enough

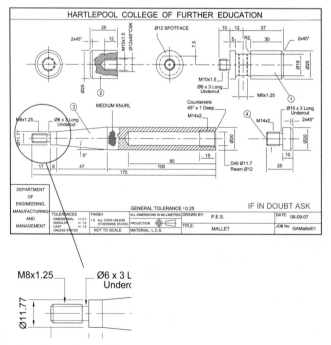

**Figure 2.14** Detailed drawing of a mallet

dimensions are provided to enable the engineer to manufacture each part. Other information is also provided, which will now be discussed.

M8 × 1.25 represents the type of thread on this part of the mallet. The 'M' means that it is metric and the 8 means that it has a diameter of 8mm. The 1.25 represents the pitch of the thread (this is the distance between the highest points of the thread).

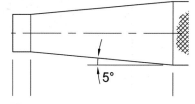

**Figure 2.15** Angles and chamfers

There are a number of words on Figure 2.14 that give the user machining instructions. These include 'drill', 'undercut', 'spotface', 'countersink (CSK)', 'ream' and 'knurl'. Machining terminology is discussed in more detail in Unit 14 (Selecting and using secondary machining techniques to remove material).

Any angles or chamfers are also included on a detailed drawing so that the product can be machined accurately (Figure 2.15).

# Exploded drawings (or exploded views)

Exploded drawings are also commonly used in engineering. These are mainly used to give customers (or manufacturing staff with less expertise of engineering drawings) an indication of how to assemble and dismantle a product. Exploded views are included when purchasing products (e.g. flat-pack furniture) or if a product needs to be maintained or repaired.

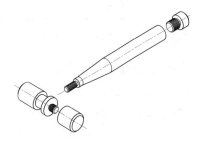

**Figure 2.16** Exploded view of a mallet

An exploded view of our mallet is shown in Figure 2.16. There is not as much detail as in the previous drawings but the assembly of the product is self-explanatory from the exploded view.

# Circuit diagrams

There are a number of different circuit diagrams used in engineering. Three of them will now be discussed: hydraulic, pneumatic and electrical wiring diagrams.

## Hydraulic-circuit diagrams

Hydraulic circuits can be represented by graphical, pictorial or block diagrams and are essential from circuit design to hydraulic trouble-shooting. Table 2.7 illustrates the elements of a basic hydraulic circuit.

| Hydraulic element | Symbol | Hydraulic element | Symbol |
|---|---|---|---|
| Reservoir (tank) | | Actuator (cylinder) | |
| Filter | | Relief valve | — |
| Pump | | Pressure line | |
| Directional control valve (DCV) | 21BAR(300PSI) | Pilot line | - - - - - - - - - - |

**Table 2.7** Elements of a basic hydraulic circuit

The valves and actuators change depending upon the type of circuit needed, but using the elements given in Table 2.7, the circuit diagram can be drawn as shown in Figure 2.17.

Note that the symbol for a hydraulic reservoir (oil tank) is repeated three times, but the actual system will only have one reservoir. The symbol is repeated to avoid too many lines being drawn back to the reservoir, which would look confusing on a more complex circuit.

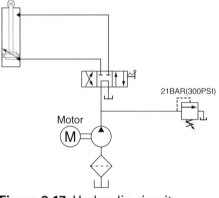

**Figure 2.17** Hydraulic-circuit diagram

The system pressure is usually indicated near the relief valve and the metric units used are bars. Some circuits may also have imperial units measured in pounds per square inch (PSI). In this case, it is 21 bar (300 PSI). If you need to convert from bars to PSI, the following formula is used:

1 bar = 14.5 PSI

If you have understood the basics of hydraulic circuitry, you will find it easier to read more complex circuits – for example, when the elements in Figure 2.17 are duplicated with more valves and actuators.

## Pneumatic-circuit diagrams

Pneumatic-circuit diagrams are very similar to hydraulic-circuit diagrams as the principles are the same (pneumatic systems use compressed air, whereas hydraulic systems use a liquid, usually oil).

A quick way to identify whether the circuit is hydraulic or pneumatic is by the symbol that represents the direction of flow (see Table 2.8).

| Circuit element | Symbol | Description |
| --- | --- | --- |
| Direction of flow (hydraulic) | ▼ | The symbol is solid |
| Direction of flow (pneumatic) | ▽ | The symbol is open |

**Table 2.8** Identifying hydraulic and pneumatic circuits

The pneumatic symbol is illustrated in Figure 2.18, where the pilot lines (the dotted lines that show the transmission of fluid pressure to control a device) indicate the direction of flow with an 'open arrow'.

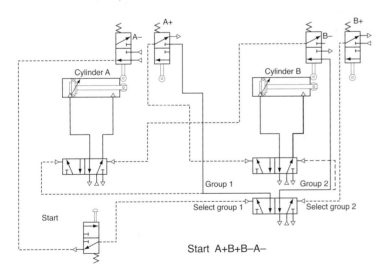

**Figure 2.18** Pneumatic- (sequence-) circuit diagram

The pneumatic circuit in Figure 2.18 is more complex than the hydraulic circuit in Figure 2.17 and is a typical example of a sequence circuit. The sequence in which the circuit operates is stated at the bottom of the circuit and indicates that cylinder A extends (A+) before cylinder B extends (B+), then cylinder B retracts (B–) before cylinder A retracts (A–). This is a common sequence circuit that uses cylinder A as a clamp with cylinder B carrying out an operation such as a stamp.

**Figure 2.19** Pneumatic air-pressure symbols

There are two ways to indicate the pneumatic power source entering the system (Figure 2.19); this indicates that there is compressed air entering at this point. This is different from a hydraulic circuit where a reservoir, pump and motor are used to indicate that hydraulic oil is entering the system at pressure.

The way in which a valve actuates (operates) depends on the system and how it is controlled. The most common methods of controlling a valve and their related symbols are shown in Table 2.9.

| Type of actuation | Symbol |
|---|---|
| Push button | |
| Lever | |
| Solenoid | |
| Pedal | |
| Roller – with spring return |  |

**Table 2.9** Methods of controlling a valve

The roller-operated valve contains a symbol on the right-hand side, which represents a spring. This indicates that the valve automatically returns to its original position once the roller is released. The spring-return symbol could be used on any of the other types in the table.

## Activity

Figure 2.20 shows a basic hydraulic circuit. The circuitry is incorrect. There are a number of mistakes involving the symbols and the names of the components need to be checked.

1. Use two sources of information to identify the components in the circuit. Make a note of your information sources.
2. Use the information to check the circuit design (has it been piped up correctly?).
3. Check that the symbols on the circuit diagram are correctly drawn.
4. List the incorrect features of this drawing from the information sources used and sketch the correct circuit diagram.

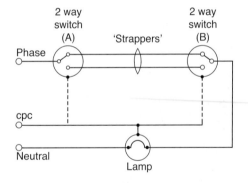

**Figure 2.20** Basic (incorrect) hydraulic circuit

## Electrical-wiring diagrams

Figure 2.21 shows a two-way lighting circuit that is used on staircases so that the downstairs light can be switched on/off upstairs, and vice versa. The diagram is quite simple to interpret as the three main conductors are identified: phase (brown), neutral (blue) and CPC (green and yellow); CPC stands for 'circuit protective conductor', otherwise known as 'earth'.

If you follow the line around from phase, it goes through switch A, through one of the two strappers (these are the two conductors between switches A and B) and switch B then lights up the lamp. The reason the lamp lights up is because a complete circuit is established. Note that the lines in the switches would normally be in black but have been shown in red to make it easier to follow the next step.

Now, if switch A or B were moved to the downward position, as in Figure 2.22, then the light would go out as there is not a complete circuit. This is typical of an electrical-wiring diagram and it contains some of the most common symbols (switches, lamp and the three main conductors).

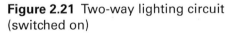
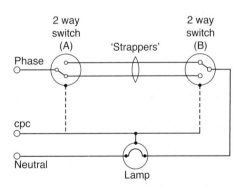

**Figure 2.21** Two-way lighting circuit (switched on)

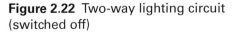

**Figure 2.22** Two-way lighting circuit (switched off)

## Pictorial (or layout) diagrams

Pictorial (or layout) diagrams can be used to show the external appearance of a circuit and look very similar to a photograph. Figure 2.23 shows a pictorial view of the previously discussed two-way lighting circuit.

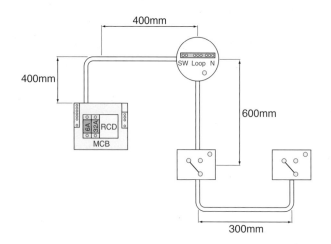

**Figure 2.23** Pictorial diagram of a two-way lighting circuit

A pictorial view can differ depending upon the company or artist/draughtsman producing it. Sometimes they are in 3D or contain very artistic drawings of components, making it even clearer to the reader.

# Graphical representations

P3

There are many ways to represent information and one of those ways is graphically. This section looks at different methods of presenting information graphically. It is likely that you will already have experience in this area.

## Sketches

Sketches are used in the early stages when designing a product. This initial stage is also referred to as a concept design and is usually developed further when accurate engineering drawings are carried out. Sketches can include dimensions and other information found on standardised drawings. It is a good idea to go around the periphery of a sketch with a darker colour so it stands out. Sketches used to be quick drawings made using a pen/pencil and paper, but now there are many computer software packages available (e.g. Google SketchUp™) that produce professional-looking sketches in a short amount of time.

Sketches are ideal for passing information on quickly if there is not enough time to complete an engineering drawing to the appropriate standards. Sketching will change from one person to another; there are no standards or rules when producing a sketch. Sometimes sketches used in industry are stored electronically and are just as important as standardised drawings. Figure 2.24 shows a sketch produced on Google SketchUp™, which could be used to show how round bar is held for marking out accurately.

## Schematic diagrams

Schematic diagrams are slightly different from the circuit and wiring diagrams that have already been discussed in this unit. They are designed to be easier to read and, therefore, you do not have to be an expert to understand them. Circuit diagrams use symbols to represent components, whereas schematic diagrams use graphical symbols that are easier to follow. A good example of a schematic diagram is the London Underground map, where each station is represented by a dot and the distance between stations is not drawn to scale. This gives the passengers a user-friendly map that is simple to follow, even though it is not drawn to scale (unlike most other maps).

## Flow diagrams

Flow diagrams are used to graphically represent a sequence of operations in a process. They consist of a series of shapes that are used to represent a part of the process. Flow diagrams are used to give the reader step-by-step instructions for activities such as trouble-shooting (finding a fault) or reducing unnecessary steps in a process. Examples of the shapes and their meanings are shown in Table 2.10.

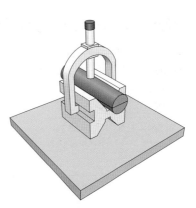

**Figure 2.24** Example of a sketch

| Shape/symbol | Meaning |
|---|---|
|  | Referred to as a terminator, this symbol represents the start or the end of a process. |
|  | This symbol means that a decision needs to be made. The chart flows in the direction of the answer. |
|  | This rectangle means that a process will take place. The process is written inside the rectangle. |
|  | The parallelogram represents data. It is also know as the input/output symbol. |
|  | This symbol represents a document. This could be a hard copy of a report, input or output. |

**Table 2.10** Flow-diagram shapes

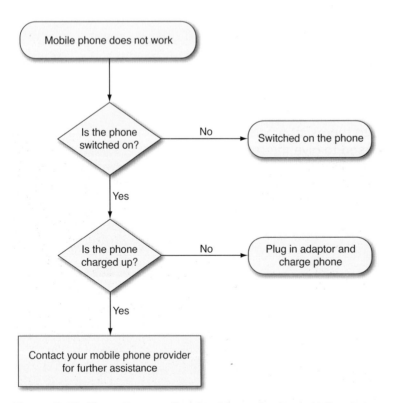

**Figure 2.25** Flow diagram for checking a faulty mobile phone

An example of a basic flow diagram is shown in Figure 2.25. This diagram goes through the steps for checking a faulty mobile phone. Note the three different shapes used in the flow diagram, which represent different activities in the process.

The shapes/symbols for a flow diagram are easily accessed on Microsoft® Word by viewing the AutoShapes/Flowchart toolbar. This is shown in Figure 2.26.

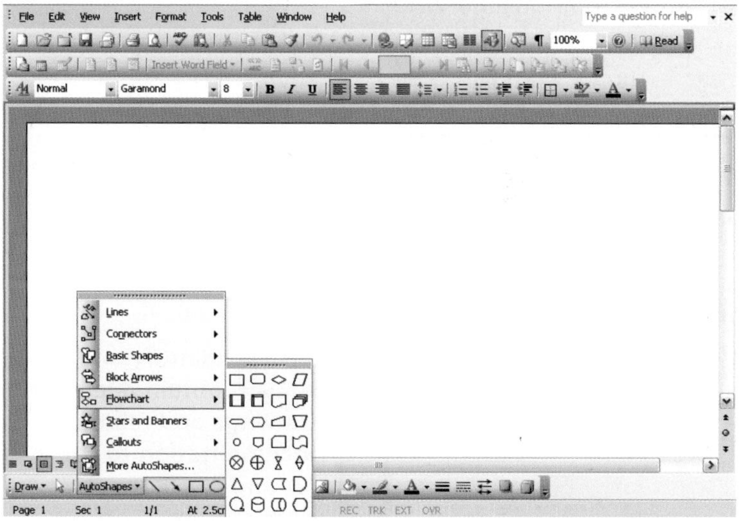

**Figure 2.26** Flow-diagram shapes in Microsoft® Word

## Physical-layout diagrams

Physical-layout diagrams can be used to show a product to prospective customers or designers before taking it to the next stage. This type of diagram will be modified at a later stage but it does give the designer something to work with, a bit like a prototype. An example of a physical-layout diagram is shown in Figure 2.27, the layout of a kitchen after the initial design stage. This gives the customer the chance to change the design to meet their requirements before an engineering drawing containing all the dimensions is drawn up for the builders.

**Figure 2.27** Physical-layout diagram

# Manufacturers' manuals

Manufacturers' manuals are an excellent source of information for most products. You will find manufacturers' manuals with mobile phones, games consoles, televisions, washing machines, microwaves, etc., all of which will include instructions for use and sometimes basic maintenance and repair notes. Detailed manufacturers' manuals for cars are usually bought separately and can be an essential source of information for car enthusiasts who carry out their own repairs.

## Activity

A pictorial sketch of a two-way electrical lighting circuit is shown in Figure 2.28. Answer the following questions.

1. Produce a circuit diagram for the two-way lighting circuit shown in Figure 2.28. Research the symbols used for electrical drawings and identify the sources of information used in this task (websites, books, etc.).
2. Produce a job card for the installation of the two-way lighting circuit (information on job cards can be found on page 73).
3. Produce an appropriate test card for the two-way lighting circuit (information on test cards can be found on page 75).

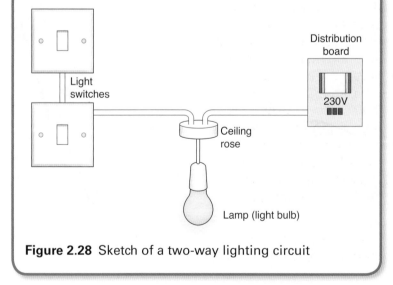

**Figure 2.28** Sketch of a two-way lighting circuit

## Make the grade

This activity will help you in achieving the following grading criterion:

**P3**  identify and obtain relevant drawings and related documentation to carry out and check own work output.

# Learning Outcome 2. Be able to use information from drawings and related documentation

## Be able to use information from work-output drawings

P4  M1  D1

We have previously examined the techniques used to interpret engineering drawings; in the following sections, we will look at how to use this documentation. When carrying out even the simplest of engineering tasks, having the correct supporting documentation is vital. For example, if we were to attempt to fabricate and assemble a simple mallet without any drawings or instructions, the task would be destined to fail, with numerous mistakes made throughout the process.

### Product-manufacture/assembly/design drawings

When a product is to be designed or assembled, a product-assembly drawing is commonly used to aid the process. As discussed on page 57, assembly drawings are used to illustrate how a product or component fits together. The diagram shows a cross-sectional view of the component and the various parts are identified using ballooned numbers. Depending on the amount of component parts, the parts list is shown either on the diagram or on a separate page. Manufacturers' part numbers and quantities may also be identified so that replacement parts may be obtained. An assembly diagram for a simple mallet (as used earlier) is shown in Figure 2.29. The assembly diagram contains no dimensional data for the individual components. This is to prevent the diagram from being swamped with too much information, thus allowing it to be read easily. Dimension details for the mallet will be found in the detailed drawings for each of the individual components. The assembly diagram may also include information that is needed to assemble the product; this could be bolt sizes, torque settings or even a specific assembly order. The overall size and weight of the assembled part may also be given; these details are of particular use when making packaging or transport arrangements.

Product-assembly drawings may also give reference to sub-assembly drawings. The sub-assembly diagram details

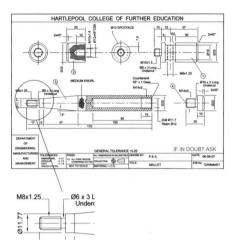

**Figure 2.29** Assembly diagram for a simple mallet

how several parts are assembled before being fitted to the main component. This type of diagram is essential when carrying out maintenance activities because it provides detailed assembly information that would be needed for a complete strip-down or overhaul of a particular component.

## Maintenance-planning documentation

Another form of engineering documentation is the planned preventative maintenance (PPM) report. Routine maintenance is essential for reducing hidden costs associated with engineering equipment. Carrying out a scheduled set of maintenance checks helps reduce machine faults and breakdowns. It improves quality and reliability and, above all else, ensures the machine is maintained to a safe working condition.

In order to record PPM activities, maintenance or facilities departments commonly use a standard PPM sheet. This record sheet serves three main purposes:

1. To act as a permanent record of the machine's maintenance requirements and provide a service history detailing each of the PPM inspections.
2. To track any trends in faults or damaged components so that new maintenance routines can be put into place to counteract these failures.
3. To be a working, live document that will be used until the machine is back to full operating condition should any repairs be needed.

A typical PPM check sheet for a hydraulic training station is shown in Figure 2.30.

| Planned Preventative Maintenance (PPM) Schedule | | | |
|---|---|---|---|
| **Description of equipment:**<br>Hydraulic Test Station | **Description of Job:**<br>Planned Maintenance Check | | **Frequency:**<br>Monthly |
| **Start date:** | **End date:** | **Location of equipment:** | **Equipment:** |
| **Tools and equipment required:** | | **Safety measures:** | |
| Socket Set | | Isolate oil supply before starting checks | |
| Rags | | Check the dump valve works correctly | |
| Allen keys | | Isolate the power to the motor | |
| Screwdrivers | | Barrier cream to be used if required | |
| Combination spanners | | | |
| Overalls | | | |
| Safety Boots | | | |
| | | | |
| | | | |
| **List of tasks/checks** | | **OK** | **NG** |
| Check for split pipes | | | |
| Check system pressure, 21 bar | | | |
| Check the oil level in reservoir | | | |
| Check the condition of the filter | | | |
| Check the pump for signs of leaks | | | |
| Check temeperature of oil | | | |
| Check component security | | | |
| Check the gauges work | | | |
| Check the solid/rigid pipe for leaks | | | |
| Check the fittings for leaks | | | |
| Clean drip tray | | | |
| Check supply lead and plug | | | |
| | | | |
| | | | |
| | | | |
| | | | |
| | | | |
| | | | |
| | | | |
| | | | |
| **Fittings and follow-ups:** | | | |
| **Completed by:** | | | **Date** |

**Figure 2.30** Planned preventative maintenance check sheet

The PPM sheet in Figure 2.30 is only part completed. This is because it is a working document, which is completed while the maintenance work is being carried out. The documentation is used in the following ways:

- Description of equipment – brief description of the equipment being checked.
- Description of job – is it a PPM, a breakdown follow-up, commissioning, etc.?
- Frequency – is it monthly, weekly or yearly?
- Location of equipment – describe where the item is located.

- Equipment ID – list the identification or serial number of the equipment.
- Tools and equipment – list all tools and equipment needed for the job.
- Safety measures – list all steps needed and any hazards.
- Tasks/checks – fully describe all checks to be completed and state if OK or no good when checked.
- Findings – if state is no good (NG), why is equipment here and is any further action needed?
- Completed by – finally sign off the PPM.

## Activity

The aim of this activity is to allow you to demonstrate an understanding of the necessary planned preventative maintenance documentation. You will carry out a simple maintenance task and complete the necessary paperwork to document your findings. The information will then be modified to enable a second maintenance task to be completed.

1. You are to carry out a one-monthly planned maintenance check on a hydraulic test station. Using the documentation shown in Figure 2.30, complete the necessary checks and fully document your findings on the maintenance report. The completed report should be clearly presented with all sections completed.

2. You are now required to carry out a second maintenance check on an electro-pneumatic test station. Clearly identify on the hydraulic-maintenance report which information is no longer suitable, is incorrect or is in need of alteration. Submit a written request explaining any information that is needed to enable the task to be carried out.

3. In response to your written request in No. 2 above, identify four sources that could be used to obtain the information required. Clearly justify each source, explaining why it could be used. Considerations should be given to validity of information, availability, ease of interpretation and the quality of the information.

### Make the grade

This activity will help you in achieving the following grading criteria:

**P4** complete all necessary production documentation related to own work output;

**M1** identify gaps or deficiencies in the information obtained that need to be resolved to enable a given task to be carried out;

**D1** justify valid solutions to meet identified gaps or deficiencies with the information obtained.

# Be able to use information from production documentation

When a product is to be manufactured, the documentation for production can typically be divided into three sections. These are as follows:

1. Work instructions – these may be in the form of job cards, standard operation procedures, modification requests or maintenance instructions.
2. Test reports – this type of report can be completed during various stages of production to track the quality and operation of the work or at the final stage of production prior to use.
3. Quality-control documentation – this documentation may be specific to an organisation. Some common quality documents are component/part inspection reports, random sampling reports of manufactured products or end-of-shift maintenance reports to monitor machine efficiency.

A typical example of each of these documents will now be examined.

## Work instructions – job cards

Job cards are completed by an employee to formally request work to be carried out. A typical job card is shown in Figure 2.31, requesting the installation of a two-way lighting circuit to be completed as part of an electrical installation exercise. However, the same job card can be used by all departments of a company. Using a job card allows the person completing the card to clearly document what work is being requested and leaves little room for errors during the work stage. It also provides the person completing the task with a useful reference, which can be referred to while the work is being carried out.

The first section of the job request indicates the unique job request number, which is required to reference and log the work when completed. The remaining information should be self-explanatory; the card is designed in such a way that a non-technical person should be able to complete the request.

The final sections of the request require the job to be authorised. This section is often completed by a supervisor or team leader. This enables the request to be assessed and a priority level assigned. The supervisor will also decide who should complete the job and if any further

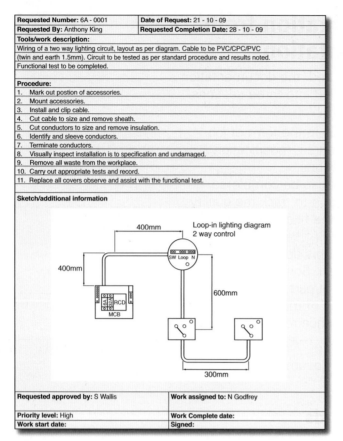

| Requested Number: 6A - 0001 | Date of Request: 21 - 10 - 09 |
|---|---|
| Requested By: Anthony King | Requested Completion Date: 28 - 10 - 09 |

**Tools/work description:**
Wiring of a two way lighting circuit, layout as per diagram. Cable to be PVC/CPC/PVC (twin and earth 1.5mm). Circuit to be tested as per standard procedure and results noted. Functional test to be completed.

**Procedure:**
1.    Mark out postion of accessories.
2.    Mount accessories.
3.    Install and clip cable.
4.    Cut cable to size and remove sheath.
5.    Cut conductors to size and remove insulation.
6.    Identify and sleeve conductors.
7.    Terminate conductors.
8.    Visually inspect installation is to specification and undamaged.
9.    Remove all waste from the workplace.
10.  Carry out appropriate tests and record.
11.  Replace all covers observe and assist with the functional test.

**Sketch/additional information**

| Requested approved by: S Wallis | Work assigned to: N Godfrey |
|---|---|
| Priority level: High | Work Complete date: |
| Work start date: | Signed: |

**Figure 2.31** Job card

information is required. Once the job is started, dates are logged and a completion date is entered before the job is signed as complete.

## Activity

**Using a blank job card, complete a request for an engineering operation. This could be a simple operation, e.g. tapping a hole or manufacturing a bracket, or something more complicated, e.g. making a drill drift.**

## Test reports

Once a task is completed, it is vital that some form of testing is carried out to verify that the aims of the task have been met. Equally important is recording this testing. The test report acts as evidence of two main points. Firstly, it provides evidence that some form of testing has been completed and it clearly indicates when this was completed and by whom. Secondly, it acts as benchmarking data, providing what is often a set of ideal results obtained when the machine or component has yet to go into production. Figure 2.32 shows a test report used by an engineer to test an electrical installation job. This report has been completed for the two-way lighting circuit and fully documents all necessary tests required prior to the job being signed off.

| Job Title: Two way lighting circuit test | | | | Date of test: 28-10-09 | | | Completed by: N Godfrey | | Signed: | | | |
|---|---|---|---|---|---|---|---|---|---|---|---|---|
| DB Circuit | Circuit Design-nation | No of Points served | Type Of Wiring | Circuit Length m | Size of circuit Conductors mm² | | Current Rating of Fuse/mcb | Circuit Continuity Ω | | Insulation Resistance MΩ | | | Polarity Check |
| | | | | | Live | cpc | | cpc | Ring | L/N | L/E | N/E | |
| 1 | light | 1 | PVC Twin & Earth | 2.5 | 1.5 | 1 | 6 amp | 0.13 | N/A | >200MΩ | >200MΩ | >200MΩ | √ |

**Figure 2.32** Test report

The top sections of the test report are completed on the day of testing by the engineer carrying out the tests. Once the test report has been completed, the document will be signed. The blue text in Figure 2.32 would be completed as the tests take place. You can see that the circuit is connected to distribution board (DB) 1 and that it is a lighting circuit. The cable is indicated as PVC twin and earth, with a circuit length of 2.5 m. The cross-sectional area of the conductor is 1.5 mm² line and neutral, and 1 mm² earth. The fuse used to protect the circuit is 6 amp. The first test to be performed is a continuity check of the circuit protective conductor; this is found to be an acceptable 0.13 Ω. A ring circuit measurement would be used if a socket ring circuit were being tested; this is not applicable in this case. An insulation resistance check is then completed between each conductor, with the result of infinity logged. Finally, the full circuit's polarity is ticked as checked and OK. This test report would then be signed and returned to the team leader, who would sign off the job as complete.

## Activity

Design a test report, listing some final tests that could be completed to check the operation of a manufactured component. This could follow on from the previous activity, e.g. checking a simple bracket or the operation of a drill drift.

# Quality-control documentation

In today's competitive marketplace, companies cannot survive without some form of quality control. This can take the form of routine checks at various stages of production or quality checks of full processes, such as a manufacturing machine. The main reasons to record quality-inspection results are as follows:

1. To highlight any trends of poor performance.
2. To record and quantify good practice or good performance.
3. To meet a company's quality-standard rating.
4. To ensure a product or machine is meeting the customer's requirements.

A quality and efficiency report is shown in Figure 2.33. This document would typically be completed at the end of a production run by a production team leader and is commonly called an overall equipment efficiency (OEE) document. This document would then be used by employees at all levels, with the aim of highlighting the losses in a process. It allows production to determine how efficient a product run was and it also allows the maintenance department to see the impact of breakdowns on a machine's performance.

There is a great deal of information in the report shown in Figure 2.33. This information has been sectioned into three main results: availability, performance and quality. The heading of the document displays the name of the machine, the ID and the date of the production run. The first section of the document displays the machine's availability; this is obtained by subtracting any planned stoppage time from the length of time the machine is planned to run (total available time). The shift breakdown records are then used to document any downtime and these total values are used to work out a percentage of how long the machine was available.

The second section examines performance efficiency and is used to determine any losses in the speed of

| | | **Body Side Press 1** | | | | |
|---|---|---|---|---|---|---|
| Equipment ID: | | Station 10 | Date: | | 22/10/2009 | |
| | | **Equipment Production Availability** | | | | |
| A: | Total available time | | | 6750 min | | |
| B: | Planned stoppage time | | | 0 min | | |
| C: | Total available time (A–B) | | | 6750 min | | |
| D: | Unplanned machine downtime taken from reports | | | | | |
| | Number of breakdowns | 1 | Total min | 40 | + | |
| | Number of adjustments of setups | 0 | Total min | 0 | + | |
| | Number of minor breakdowns | 1 | Total min | 10 | + | 50 min |
| E: | Operating time (C–D) | | | 6700 min | | |
| F: | Equipment availability F = (E/C × 100) | | | 99.30% | | |
| | | **Performance Efficiency** | | | | |
| G: | Total Parts produced | | | 800 parts | | |
| H: | Target cycle time | | | 8 min per part | | |
| I: | Performance efficiency | | | | | |
| | I = ((H × G) / E × 100) | | | 800 parts | | |
| | | **Quality Rate** | | | | |
| J: | Total defects (Rework and Scrap) | | | 10 parts | | |
| K: | Quality rate | | | | | |
| | K = ((G – J) / G × 100) | | | 10 parts | | |
| Overall Equipment Efficiency Rate | | | | | | |
| | | | | | | |
| OEE = F × I × K / 1000 | | | | 93.60% | | |
| Target 85% | | | | | | |

**Figure 2.33** Overall equipment efficiency report

production. The report states the total parts produced and the speed at which these parts should be manufactured in an ideal production run. The values of the actual production cycle time against the target cycle time produce a value to show the effectiveness of the production cycle.

The third section of the document examines quality rate. The quality of the parts produced is logged; this includes any parts that fell below the required quality level but were subsequently repaired. This value is then used to calculate the percentage of acceptable quality parts that were produced by a machine.

The final step is to produce the machine's OEE rate. This is calculated using the values of availability, performance and quality to produce a final percentage. The value achieved in this case is 93.6 per cent, which is 8.6 per cent above target, indicating that the machine is running efficiently in all areas.

## ❝ Team Talk

Aisha: **'Is it possible for a machine to produce an OEE of above 100 per cent?'**
Steve: **'No. If it does, then one of the values has been incorrectly entered. It is important to be honest with the information entered. It would be quite easy to miss out a defective part or put in a low cycle time, but this would invalidate the results, rendering the entire process useless.'** ❞

### Make the grade

The next activity will help you in achieving the following grading criterion:

 **P4** complete all necessary production documentation related to own work output.

## Activity

An OEE report is required to determine the efficiency of a manufacturing process. For this activity, the process is the manufacturing of simple paper aeroplanes. As a class, complete the following activity:

- Decide on a paper aeroplane design and issue identical manufacturing diagrams/ instructions to at least four production workers.
- As a group, determine and record the following parameters on a blank OEE form:
    - total available production time;
    - planned stoppage time;
    - target cycle time.
- Start production of your paper aeroplanes, recording any unplanned stoppage time, e.g. for poor-quality reworks, loss of resources, damage, etc.
- Once production time is finished, complete the OEE sheet with the necessary data and calculate your overall OEE.

1. Was your OEE above the 85 per cent target? If not, what part of the process let you down?
2. How could OEE be improved without altering the production time, planned stoppage time or cycle time?
3. What effect would inaccurate/poor-quality assembly diagrams have on OEE?

# Drawing and document care and control

An engineering drawing or technical document exists because the information presented on it has value to someone in the engineering organisation. Therefore, it goes without saying that if this document is misplaced, inaccurate, damaged or unusable, there will be some form of impact on the organisation. This section looks at various methods of document care and control employed by companies to safeguard engineering drawings and production documentation.

## Location and storage

Drawings and documents can be presented in two forms: paper and electronic copies. The storage of both will now be examined.

### Paper documents

Original master documents, such as technical drawings, must be stored in a secure storage cabinet. The size of the cabinet depends on the size of the document, and its two main functions are to keep the drawing away from any oil, dirt or contaminants and to allow some form of filing system to be used. Paper copies of these documents are often stored in more accessible locations. For example, an electrical wiring diagram is often stored in the filing cabinet of the machine it relates to. This gives the maintenance engineer quick and easy access when fault-finding on a machine.

Copies of OEE reports are often placed in display boards around the factory to show how efficiently a machine is running. Commonly used assembly diagrams can be laminated or stored in plastic wallets, allowing them to be used freely by production staff without the risk of contamination.

**Figure 2.34** Filing cabinet

## Activity

**Your class notes are vital documents that need to be referred to when completing assessments and end tests. As a class, list several headings that could be used to assess the quality of storage of your class notes. Some examples are listed below:**

- **What is the storage method?**
- **Are lessons sectioned?**
- **Are handouts/notes dated?**
- **Are there assessment/criteria references?**
- **What is the condition of the notes?**
- **Are all notes present?**

**Carry out a notes scrutiny exercise in groups and present your findings. To finish this activity, decide on an ideal storage method that addresses any problems found.**

## Electronic documents

Most engineering drawings and production documents are now compiled using some form of computer-aided drawing (CAD) package (as covered in Unit 10). This has the distinct advantage of allowing the master document to be stored easily in a secure location. All diagrams associated with a particular part are commonly stored in a complete file, which can then be copied to a back-up CD or memory stick. Production documentation is stored in the same way and often filed according to the machine and the week or month of the production run.

## Security control

Many larger companies will have some form of drawing office, which will be responsible for the storage and security of the master engineering drawings. In smaller organisations, the security will often involve the simple but effective method of a locked filing cabinet. The keys for this cabinet would then be held by a nominated employee. The purpose of using a drawing office or nominated employee is to provide only one access point to the master documentation. The person responsible could then provide a set-up similar to a library, whereby master documents are requested and booked out on loan. A record of who has the document, when it was borrowed and when it must be returned could be kept in order to keep it secure at all times. Electronic documents could be stored with password protection and simply printed out as required. Companies will often adopt an access-and-return procedure for using master diagrams; this may take the form of submitting a written request signed by a supervisor, or a formal email. The request would state the specific document required, why it is needed and the issue/return dates.

## Report discrepancies in data and documents

Due to the amount of information and detail presented on engineering drawings and production documentation, it is not uncommon for mistakes to be made. It is very important that when a discrepancy is discovered, a procedure is in place to report it. As previously stated, the engineering drawing used on the shop floor is never the original; therefore, if a dimension is incorrect and altered on this copy without changing the master, any further copies of the master will be incorrect. In most cases, to encourage people to report discrepancies, the procedure for doing so is as simple as possible. It is often just a case of marking up the copied diagram with the alterations and completing a brief amendments request. A typical request form is shown in Figure 2.35.

| Diagram/document Modification Request | | |
|---|---|---|
| Title of diagram/document | | |
| Drawing/document number | | |
| Date request made | | Requested by |
| Details of request | | |
| | | |
| Authorise | | Completed by |
| Date | | Date |

**Figure 2.35** Diagram/document modification request form

This request would then be submitted to the drawing office or person in charge of the engineering documentation. It is important that any modifications are reported so that the original diagram is kept up-to-date.

# Physical handling of documents

To ensure a document is efficiently used, it is important that it is presented and handled with care and attention. Damage, graffiti and contamination of a master engineering drawing can be avoided by adopting the security and storage measures previously discussed. For working documents, simple measures such as laminating the diagram, storing it in plastic and providing file storage space on the shop floor can help keep it in good condition. It is highly unlikely that a working diagram will stay presentable forever and that is why the original should never be used.

The simple method of correctly folding a document so that the title block is always visible can prevent numerous diagrams being removed from a file and opened out to locate the diagram required. Wherever possible, diagrams should always be folded to A4 size; this allows them to be stored together and prevents the need for numerous files containing various diagrams.

# Document control

All engineering drawings contain key information, as discussed earlier in this unit. Three vital pieces of information that are key to the control of the document are the issue dates, amendment dates and pattern/part numbers.

## Issue and amendment dates

The issue date is the date at which the drawing or report was completed and is commonly stated when referring to a diagram. The issue date is important because it is often a unique feature when combined with a part/pattern number. If a diagram is modified, the original date should remain on the diagram but a modified date should also be recorded. This modified date is called the amendment date. It is also common practice to record a revision with a number or letter after the drawing number. For example, the drawing number 116/C or 116/2 would indicate revision C or amendment 2. This shows how many modifications have been made. Once a revised diagram or report is produced, all preceding diagrams should be destroyed.

## Part/pattern numbers

The part or pattern numbers are very specific and should refer to one diagram only. They often have a meaning to the company that produced the diagram, the part itself or the company for whom the diagram was produced. They are commonly used when referring to a diagram, for location, storage, and modification requests. For example, 'AK – HART – Mal178' contains four pieces of information: 'AK' is the person who completed the diagram, Anthony King; 'HART' is the company for whom the diagram was produced, Hartlepool Engineering; 'Mal' is the name of the part, mallet; and '178' is the part number relevant to Hartlepool Engineering.

## Reporting of loss or damage

If a master drawing or document becomes lost or damaged, then, providing an electronic copy exists, it is simply a case of requesting another printout. This may be requested or reported verbally but it is often necessary to send a written request to the drawing office via email. It is more serious if the original is lost and no copy exists. This would still need to be reported to a member of the drawing office, who may have to commission another drawing or report. The important point is that the loss is reported so that immediate action can take place and countermeasures be taken while the reports are being obtained.

### Make the grade

The next activity will help you in achieving the following grading criteria:

**P5** describe the care and control procedures for the drawings and related documentation used when carrying out and checking own work output;

**M2** identify improvements in the care and control procedures used for drawings and related documentation.

## Activity

The aim of this activity is to allow you to demonstrate an understanding of the necessary care and control procedures to be taken when using documentation to assist in carrying out your own work. You have already completed a simple maintenance task and completed the necessary paperwork to document your findings. The care and control procedures used to handle this documentation will be considered in this activity.

1. You have previously carried out a one-monthly planned maintenance check on the hydraulic test station and fully documented your findings on the maintenance report. Describe the care and control procedures that should be in place for the drawings and documentation that were used in this task.

2. Identify improvements that could be made to help care for and control the drawings and documents you used.

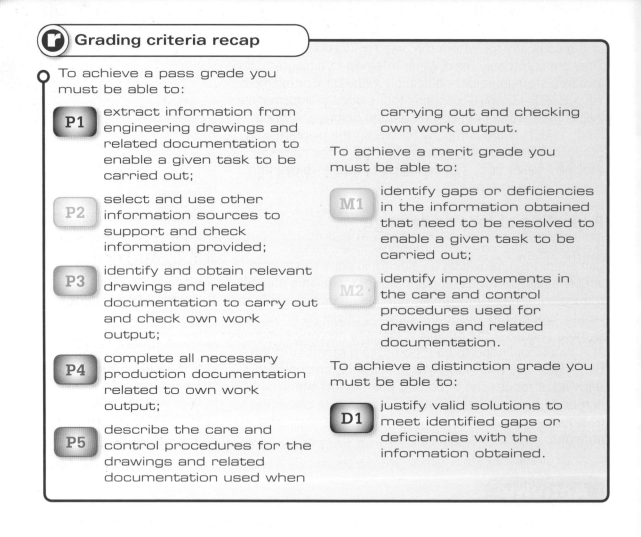

## Grading criteria recap

To achieve a pass grade you must be able to:

**P1** extract information from engineering drawings and related documentation to enable a given task to be carried out;

**P2** select and use other information sources to support and check information provided;

**P3** identify and obtain relevant drawings and related documentation to carry out and check own work output;

**P4** complete all necessary production documentation related to own work output;

**P5** describe the care and control procedures for the drawings and related documentation used when carrying out and checking own work output.

To achieve a merit grade you must be able to:

**M1** identify gaps or deficiencies in the information obtained that need to be resolved to enable a given task to be carried out;

**M2** identify improvements in the care and control procedures used for drawings and related documentation.

To achieve a distinction grade you must be able to:

**D1** justify valid solutions to meet identified gaps or deficiencies with the information obtained.

# Unit 3
## Mathematics for engineering technicians

## Introduction to the unit

In many ways, mathematics and engineering are so closely related that they are practically inseparable. Just about every aspect of engineering requires maths of one sort or another. When you carry out a basic drawing or a marking-out exercise, then at the very least you need to use addition and subtraction; when designing a new workshop layout, you need to consider the area and volume of the space available; when operating machining tools, you need to understand standard form and scientific notation; and when working with fluid systems or designing electric circuits, you most certainly need to rearrange (or transpose) equations.

In other ways, a sound understanding of maths can help you enormously in everyday life: from understanding your tax contributions and organising the best mortgage on a new home, to estimating the amount each person should pay towards a restaurant bill.

You will, of course, have studied mathematics at school, but this unit will show you how to apply maths to real engineering problems – it is certainly not maths for the sake of just doing maths!

Figure 3.1

### Learning Outcomes

By the end of this unit you should:

- be able to use arithmetic, algebraic and graphical methods to solve engineering problems;
- be able to use mensuration and trigonometry to solve engineering problems.

# Grading criteria

| To achieve a pass grade you must be able to: | To achieve a merit grade you must be able to: | To achieve a distinction grade you must be able to: |
| --- | --- | --- |
| **P1** use arithmetic methods to evaluate two engineering problems, ensuring answers are reasonable | **M1** transpose and evaluate complex formulae | **D1** transpose and evaluate combined formulae |
| **P2** use algebraic methods to transpose and evaluate simple formulae | **M2** identify the data required and determine the area of two compound shapes | **D2** carry out chained calculations using an electronic calculator |
| **P3** plot a graph for linear and non-linear relationships from given data | **M3** identify the data required and determine the volume of two compound solid bodies | |
| **P4** determine the area of two regular shapes from given data | **M4** use trigonometry to solve complex shapes | |
| **P5** determine the volume of two regular solid bodies from given data | | |
| **P6** solve right-angled triangles for angles and lengths of sides using basic Pythagoras's theorem, sine, cosine and tangent functions | | |

# Learning Outcome 1.  Be able to use arithmetic, algebraic and graphical methods to solve engineering problems

## The basics of BODMAS

D2

To quickly review some basics that you will have studied at school, Table 3.1 illustrates the fundamental symbols used in engineering mathematics.

| Maths symbol | Meaning | Example |
|---|---|---|
| + | Addition: two or more quantities are combined to find the sum total. | 10 + 10 |
| − | Subtraction: a quantity is taken from another quantity. | 10 − 10 |
| × | Multiplication: two or more quantities are 'times' together and therefore combined. Other symbols commonly used for multiplication include a dot (.) or an asterisk (*). | 10 × 10 |
| ÷ | Division: the inverse (or opposite) of multiplication, where a quantity is 'split' by another quantity. Very often a forward slash (/) is used to denote the division bar. | 10 ÷ 10 or $\frac{10}{10}$ |
| ( ) | Brackets: used to separate quantities in larger calculations. | 10 (10 − 10) + 10 |
| = | Equal: used to denote the outcome of a sum or to denote that two sides of a sum will balance. | 10 + (10 − 10) = 10 |
| . | Decimal point: used to show in more detail the size of a quantity. | 10.10 |
| $A^2$ | Squared: when a quantity is multiplied by itself to get a new number | $10^2 = 10 \times 10 = 100$ |
| √ | Square root: the inverse of squared. | $\sqrt{100} = 10$ |
| π | Pi: commonly used in mathematics for calculations concerning circle area and circumference. It is simply a constant number: 3.14159. | $\pi r^2$ |
| < | Less than: the quantity on the left of the symbol is less than the quantity on the right. | 9 < 10 |
| > | More than: the quantity on the left of the symbol is more than the quantity on the right. | 10 > 9 |
| ≠ | Not equal: the two sides of a calculation do not balance. | 10 + 10 ≠ 10 × 10 |

**Table 3.1**  Basic symbols used in engineering mathematics

 **Team Talk**

Aisha: **'What is BODMAS?'**
Steve: **'It's just an acronym.'**
Aisha: **'What's an acronym?'**
Steve: **'It's when the letters stand for something: FA stands for Football Association, RFU stands for Rugby Football Union – both are common acronyms.'**

Note: a small calculator is shown throughout the unit to illustrate the use of a scientific calculator.

When calculating a quantity, equation or formula that contains a combination of the symbols in the previous table, it is important to follow the rule of BODMAS (sometimes called BIDMAS). This term helps you remember what to do and in what order:

**B** = Brackets (work out the brackets first)
**O** = Order (work out the powers or square roots)
**D** = Division (work out any division)
**M** = Multiplication (work out the multiplication)
**A** = Addition (work out the addition)
**S** = Subtraction (finally, carry out any subtraction)

Sometimes the 'O' is replaced by an 'I', which means 'indices' – it does not change the meaning.

## Worked Example

**Example 1:**

If you want to work out:

$$(12 \times 15) \times 10, \text{ then:}$$

First work out the brackets:

$$12 \times 15 = 180 \qquad \boxed{\textbf{BODMAS}}$$

Then multiply by the 10:

$$180 \times 10 = 1800 \qquad \boxed{\textbf{BODMAS}}$$

**Example 2:**

$$8 + 10(15 \times 10) - 9$$

First work out the brackets:

$$15 \times 10 = 150 \qquad \boxed{\textbf{BODMAS}}$$

Then multiply by the 10 (note that when a number is directly next to the brackets it means multiply – it is a kind of shorthand, which is why the 'x' symbol is not shown):

$$150 \times 10 = 1500$$

| BODMAS |

Now add the 8:

$$1500 + 8 = 1508$$

| BODMAS |

And finally subtract the 9:

$$1508 - 9 = 1499$$

| BODMAS |

**Example 3:**

$$(10 \times 10)12^2 \div 2$$

First work out the brackets:

$$(10 \times 10) = 100$$

| BODMAS |

Now work out the order:

$$12^2 = 12 \times 12 = 144$$

| BODMAS |

When we arrive at the division, we do:

$$144 \div 2 = 72$$

| BODMAS |

And finally we multiply by 100:

$$72 \times 100 = 7200$$

| BODMAS |

Many calculators have the rule of BODMAS programmed in as part of their function. In other words, as long as you enter the calculation carefully, it will provide you with the correct answer.

## Worked Example

**Example:**

We worked out that:

$$(10 \times 10)12^2 \div 2 = 7200$$

We can enter the calculation in the original order by pressing the following keys carefully:

    [ ( ]    [10]    [×]    [10]    [ ) ]    [×]    [12]    [$x^2$]    [÷]    [2]    [=]

Alternatively, you could enter:

    [12]    [$x^2$]    [ ( ]    [10]    [×]    [10]    [ ) ]    [÷]    [2]    [=]

## Activity

**Have a go at the following calculations:**

1. **(10 + 10) × 46(78 + 98) − 10**
2. **78 + 98(45 − 14)**
3. **(10 ÷ 2) × 6 − 6**
4. **252 × 79(12 + 12) − 56**

**You can find the answers in the back of the book.**

## Make the grade

It is essential that you can use a calculator in one continuous operation to achieve the following grading criterion:

**D2**  **carry out chained calculations using an electronic calculator.**

Your teacher or college tutor will be able to show and support you in doing this and, in some cases, will need to witness that you can do it.

There will be a number of tips throughout this unit to help you get started, but calculators come in a range of models from a number of different manufacturers. All scientific calculators have the basic features discussed in this unit, but the way you use the functions will inevitably vary from one calculator to another – factory default settings may also cause some confusion. We have used a very popular Casio model in our illustrations. However, you could use the calculator on Windows® and even some mobile phones and media players now have scientific calculators as applications ('apps').

If you need help using your calculator, seek guidance from your teacher/tutor. The chances are they will have seen the calculator before or will be able to adapt to a new model relatively quickly.

# Basics: decimals, ratio, proportion and percentage

Now that we have introduced BODMAS, it is worthwhile revising a number of topics that you will have previously covered at school or college.

An engineer will never concentrate solely on specific technical problems involving complex mathematics; inevitably, from time to time, all engineers will be required to report progress to senior management or present financial or operational targets. Engineers often need to take complex information and adapt it so that it can be easily understood by interested parties and stakeholders (such as consumers, customers, shareholders and suppliers).

## Fractions

In school, fractions have been part of the mathematics curriculum for years.

The top number in a fraction is called the numerator and the bottom number is a denominator. Hence, in the example below, 12 is the numerator and 36 is the denominator:

$$\frac{12}{36}$$

**Figure 3.2** Fractions

You will also recall that, to simplify a fraction to an equivalent, you need to reduce by a factor that is cleanly divisible by the numerator and the denominator.

By dividing by a factor of 2, we get:

$$\frac{6}{18}$$

By dividing this by 3, we get:

$$\frac{2}{6}$$

And dividing this by 2, we get:

$$\frac{1}{3}$$

This is said to be the simplest form of the fraction.

Of course, we could have simply divided our original fraction ($\frac{12}{36}$) by a factor of 12 to get $\frac{1}{3}$.

## Ratios

Ratios are fundamentally very similar to fractions; they are just presented differently.

If an engineering company has a total of 48 staff, 12 male and 36 female, then the ratio between males and females would be expressed as 12:36, or 1:3 if we apply the same simplification as for the previous fraction. The ratio of females to males would, therefore, be 36:12 (or 3:1).

Ratios are often used to state the scale on engineering and technical drawings. Drawings can represent very large objects (such as aircraft) or very small objects (such as electronic components). In both cases, a scale is used.

A scale of 1:1 means that the size of the object on the drawing is the same size as the real object. Table 3.2 describes a variety of scales that might be used on engineering drawings.

| Scale | Drawing size | Object size | Description |
|-------|------|------|-------------|
| 1:1 | 1 | 1 | The drawing and the object are the same size. |
| 1:2 | 1 | 2 | The object is ×2 the size of the drawing. |
| 1:5 | 1 | 5 | The object is ×5 the size of the drawing. |
| 1:10 | 1 | 10 | The object is ×10 the size of the drawing. |
| 2:1 | 2 | 1 | The object is $\frac{1}{2}$ (0.5) the size of the drawing. |
| 5:1 | 5 | 1 | The object is $\frac{1}{5}$ (0.2) the size of the drawing. |
| 10:1 | 10 | 1 | The object is $\frac{1}{10}$ (0.1) the size of the drawing. |

**Table 3.2** Description of scales

## Getting a decimal from a fraction

The quickest method for converting a fraction to a decimal is to put it in your calculator.

To work out $\frac{12}{36}$ as a decimal:

Press the [12] key

Press the [÷] key

Press the [36] key

Press the [=] key to show the answer [0.3333333333]

## Decimal places

The number presented on the calculator LCD is 0.3333333333; this is referred to as 0.3 recurring.

Engineers often need to round numbers to a certain number of decimal places (normally two or three, depending on the number).

- Shown to one decimal place, the number would be 0.3.
- Shown to two decimal places, the number would be 0.33.
- Shown to three decimal places, the number would be 0.333.

If the number to be rounded to two decimal places is 0.759801, then we need to consider the third number after the decimal place.

| 0 | . | 7 | 5 | 9 | 8 | 0 | 1 |
|---|---|---|---|---|---|---|---|

In this example, the third number after the decimal place is 9. The rule to follow is that, if the number is 0, 1, 2, 3 or 4, we round down. Numbers from 5 to 9, we round up.

As the number in the third place after the decimal point is 9, we round the number up. Therefore, the number becomes:

| 0 | . | 7 | 6 |
|---|---|---|---|

If the number is:

| 0 | . | 7 | 5 | 4 | 8 | 0 | 1 |
|---|---|---|---|---|---|---|---|

Then we would round down to:

| 0 | . | 7 | 5 |
|---|---|---|---|

## Activity

**Express the following numbers to three decimal places:**

1. 0.11111
2. 0.125889
3. 0.25689
4. 22.36548
5. 125.3692

# Percentage

The term 'percentage' also appears with some regularity in engineering.

In a production meeting, a team leader or manager may frequently quote efficiency rates; for example, 'We have a press tool down today, so at best we can hope for 80 per cent efficiency.' This means that, due to a breakdown of a machine tool, the plant can only operate at 80 per cent of its total capacity. If the plant can produce 1,000 components a day at 100 per cent efficiency, then today they can only expect to produce 800 components.

Percentages are simply decimal values expressed in a different format. Table 3.3 shows a number of examples (to move a decimal value to a percentage, you simply multiply by 100).

| Fraction | Ratio | Decimal | Percentage |
|---|---|---|---|
| $\frac{1}{1}$ | 1:1 | 1.0 | 100 |
| $\frac{1}{2}$ | 1:2 | 0.5 | 50 |
| $\frac{3}{10}$ | 3:10 | 0.3 | 30 |
| $\frac{1}{10}$ | 1:10 | 0.1 | 10 |
| $\frac{1}{25}$ | 1:25 | 0.04 | 4 |
| $\frac{1}{50}$ | 1:50 | 0.02 | 2 |
| $\frac{1}{100}$ | 1:100 | 0.01 | 1 |

**Table 3.3** Percentages

## Worked Example

A production company produces 625 components in a day. When working at 100 per cent efficiency, it can produce 1,000 components. What is its efficiency as a fraction, ratio, decimal and percentage?

The solutions are shown in Table 3.4.

| | | |
|---|---|---|
| **As a fraction** | $\frac{625}{1000}$ | This can be reduced to its simplest form: $\frac{5}{8}$ |
| **As a ratio** | 625:1000 | This can be simplified in the same way to 5:8 |
| **As a decimal** | 0.625 | This can be worked out on a calculator by keying in [625] [÷] [1000] or [5] [÷] [8] |
| **As a percentage** | 62.5% | Simply multiply the decimal by 100: $0.625 \times 100 = 62.5\%$ |

**Table 3.4** Fractions, ratios, decimals and percentages

## Activity

A resistor is a device used to reduce the electrical current flowing into a circuit or component. The coloured bands on the resistors are used to show a) the size of resistance in ohms, and b) the tolerance.

Common resistors are supplied with a tolerance of 1, 2, 5 or 10 per cent. More expensive resistors have a tighter tolerance of 0.5, 0.25 or 0.1 per cent.

Based on the tolerance and resistor value given, complete Table 3.5 with the upper and lower resistance of each component:

**Figure 3.3** Resistors

| Resistor value ($\Omega$) | Tolerance (%) | Lower limit ($\Omega$) | Upper limit ($\Omega$) |
|---|---|---|---|
| 10 | 20 | 8 | 12 |
| 25 | 5 | 23.75 | |
| 50 | 1 | | 50.5 |
| 100 | 10 | | |
| 1,000 | 0.5 | | 1,005 |
| 12,000 | 0.1 | 11,988 | |

**Table 3.5** Upper and lower resistance levels

# Standard form (scientific notation)

D2

In this section, we will introduce standard form, sometimes called scientific notation.

According to wikipedia.org, the mass of the Earth is approximately $5.97 \times 10^{24}$ kg. The number in full would be 5,970,000,000,000,000,000,000,000 kg.

Scientific notation is used to simplify very large numbers (like the mass of the Earth) or very small numbers (such as capacitance on electronic circuit boards). Scientific notation uses a decimal number followed by a power of 10; we have already used an example on this page: $5.97 \times 10^{24}$.

All numbers can be expressed in this way, using one number before the decimal point and two numbers after (two decimal places). A few examples are shown in Table 3.6.

| Example | Actual number | Scientific notation (standard form) |
|---|---|---|
| Approximate engine speed of F1 car (r.p.m.) | 20,000 | $2.00 \times 10^4$ |
| Approximate distance from London to Paris (km) | 212 | $2.12 \times 10^2$ |
| Mass of the Moon (kg) | 73,600,000,000,000,000,000,000 | $7.36 \times 10^{22}$ |
| A 2 microfarad capacitor | 0.000002 | $2.00 \times 10^{-6}$ |

**Table 3.6** Examples of scientific notation

Some caution does need to be exercised when using scientific notation in an engineering context as many values tend to be rounded to $10^3$, $10^6$, $10^9$, $10^{-3}$, $10^{-6}$ or $10^{-9}$ etc. This is to allow for prefixes and units to be applied and is discussed in more detail in Unit 4 (Applied electrical and mechanical science for engineering).

To enter a power of 10 into your calculator, you need to use the [×10ˣ] key – sometimes it is shown as [EXP], an abbreviation for 'exponent'.

For example, this is how you enter the mass of the Moon: $7.36 \times 10^{22}$
First enter your number: [7]    [.]    [3]    [6]
Now press [×10ˣ]
Enter the power: [22]
Press [=] to complete the entry and display the scientific notation on the calculator LCD.

A smaller number may be presented on the calculator LCD in its actual number straightaway; for example, $1.12 \times 10^6$ may be displayed as 1120000.

Another handy key is [ENG], which allows you to move easily between the different engineering notations for a number.

For example, enter:
1120000

Now press the [ENG] key once and it will display the number in the most appropriate engineering form: $1.12 \times 10^6$.

Press it again and it will offer $1120 \times 10^3$.

If you press [SHIFT] and then [ENG], it goes back to $1.12 \times 10^6$.

## Make the grade

Engineering form is discussed in more detail in Unit 4 (Applied electrical and mechanical science for engineering). The next activity will help you in achieving the following grading criterion:

**D2** carry out chained calculations using an electronic calculator.

## Activity

Enter the following scientific notation into your calculator and then use the [ENG] key to view the numbers written out in full:

1. $8.56 \times 10^8$
2. $9.81 \times 10^0$
3. $1.11 \times 10^3$
4. $5.56 \times 10^{12}$

Enter the following numbers in full and use the [ENG] key to display the numbers in their engineering form:

1. 1896478
2. 256895412
3. 11245
4. 125

Find the answers to the following calculations by entering them into your calculator in one operation:

1. $12 + 78 + 957 + 1.25 \times 10^3$
2. $(10 \times 10) \times 56 \times 1.25 \times 10^3$

# Transposition (rearranging formulae)

P2

Transposition of formulae (or rearranging formulae) is one of the most widely used mathematical methods in use today – you will have transposed a formula many times before and, in most cases, never even noticed you were doing it.

For example, if you went to a supermarket with £10 and worked out that you could afford a DVD priced at £9 and then expected £1 change, then you would have carried out a transposition:

Money in your pocket = DVD + change

$$£10 = £9 + £1$$

**Figure 3.4**

The formula can be rearranged as follows:

Money in your pocket − DVD = change

$$£10 − £9 = £1$$

Another supermarket example using multiplication and division is the purchase of multiple items of the same value. If you purchased ten cans of an energy drink costing £1 each, you would expect to pay £10.

The formula is:

Total cost = quantity (of cans) × unit cost (of cans)

£10 = 10 × £1

You will apply transposition if you have £10 in your pocket: you have checked the price of the cans at £1 each and have quickly calculated how many you can afford to purchase – the most basic mathematics tell you that you can afford ten cans, but the formula can be rearranged as:

$$\frac{\text{Total cost}}{\text{unit cost}} = \text{quantity}$$

$$\frac{£10}{£1} = 10 \text{ cans}$$

Both are simple examples, but they do hold up mathematically and the rules of transposition can be applied to any balanced formula. Transposition is widely used in engineering when a number of *known variables* are rearranged in a standard formula to find an *unknown variable*. Common examples (linked to engineering science) would include the use of transposition to determine mass, force (and weight), resistance, voltage, current, power, work, acceleration, velocity, distance travelled, flux and magnetic flux density.

To get you going in the right direction, remember the following:

*Try not to worry too much about the letters and symbols used in a formula. It does not matter if the formula uses the English, Greek or Roman alphabet; the letters and symbols are all used as simple abbreviations.*

In our previous example, we could have described it as:

$$\frac{\text{TC}}{\text{UC}} = \text{Q}$$

Where TC = total cost; UC = unit cost; Q = quantity

It may look slightly more technical but it still represents the amount of cans you can get for ten quid!

*The second point to remember is that the formula must always balance. It is a bit like having a set of measuring scales that must always balance – if you are transposing correctly, then the numbers and symbols on each side will always be equal (or, in reality, add up to the same value).*

$10 + 30 + 40 = 50 + 5 + 25$     (both sides sum to 80)

Rearranging could give:

$10 + 30 + 40 - 5 = 50 + 25$     (both sides sum to 75)

Rearranging could also give:

$10 + 30 + 40 - 5 - 25 = 50$     (both sides sum to 50)

$10 + 30 + 40 + 5 + 25 = 50$

*The third point to remember is that you must always be fair to both sides to ensure they balance. If you add a value to one side, you must do the same to the other; if you subtract a value on one side, then you must do the same to the other side; if you multiply, divide, square or square root a value on one side, you must do the same to the other.*

## Worked Example

**Example 1:**

$$A + B + C = D + E$$

If we wanted to make D the subject (the variable we don't yet know, but want to know), then we need to move E to the other side.

We do this by subtracting E from both sides:

$$A + B + C = D + E$$

becomes:

$$A + B + C - E = D + E - E$$

What happens next is the key to how transposition works, i.e. how parts of a formula can be rearranged to get the value of interest.

The (+ E) and (– E) simply cancel each other out on the right-hand side (RHS) – this is because they would always equal zero, e.g. $+10 - 10 = 0$.

This leaves:

$$A + B + C - E = D$$

## Activity

**Have a go at the following formula to make x the subject instead of y:**

$$y = x + 10$$

## Worked Example

**Example 2:**

$$A \times B = C \times D$$

We will rearrange the formula to make C the subject. We can do this by dividing both sides by D. (Remember, we have to be fair to both sides of the formula – the scales have to balance!)

$$A \times \frac{B}{D} = C \times \frac{D}{D}$$

On the RHS, D is both a multiplier (above the line) and a divider (under the line). Therefore, it simply equals 1 on the RHS. As a check, if C = 2 and D = 10, then $C \times \frac{D}{D} = 2 \times \frac{10}{10} = 2$ (the original value of C).

D stays on the left-hand side (LHS) as there is not another D there to simplify further.

This leaves:

$$A \times \frac{B}{D} = C$$

**Example 3:**

This third example illustrates what to do if the formula is a little more complex:

$$A \times B \times C \times D = E - F$$

If we were to make F the subject, then it would be tempting to move E to the LHS by subtracting E from both sides.

This would become:

$$A \times B \times C \times D - E = F \text{ (THIS WOULD BE WRONG)}$$

The reason is that F is actually a negative value. The formula would actually read:

$$A \times B \times C \times D - E = -F$$

To transpose this formula correctly, first we need to make F positive. We can do this by moving F from the RHS by the addition of F to both sides:

$$A \times B \times C \times D = E - F \text{ (we first add F to both sides)}$$

$$A \times B \times C \times D + F = E - F + F$$

As the Fs on the RHS equal zero, the formula can be simplified:

$$(A \times B \times C \times D) + F = E$$

It is then a case of moving the other variables to the RHS by subtracting $(A \times B \times C \times D)$ from both sides – yet again, the scales must balance!

$$(A \times B \times C \times D) - (A \times B \times C \times D) + F = E - A \times B \times C \times D$$

The LHS can be simplefied to leave:

$$F = E - A \times B \times C \times D$$

You could use the following numbers to check, if you are still unsure:

$$A = 1, B = 2, C = 3, D = 4, E = 48, F = 24$$

**Example 4:**

The final example illustrates the care required when using multiplication and division:

$$A \times B \times C = D \times \frac{E}{F}$$

If we wanted to make F the subject, then we would first need to move it above the division bar (making it a numerator rather than a denominator).

We do this we multiplying both sides by F:

$$A \times B \times C \times F = D \times E \times \frac{F}{F}$$

The Fs on the RHS equal 1 so the formula can be simplified:

$$A \times B \times C \times F = D \times E$$

To now make F the subject, we must divide both sides by $(A \times B \times C)$:

$$F = D \times \frac{E}{A} \times B \times C$$

## Activity

Make the letter in brackets the subject of the following formulae:

1. $A + b + c + D = E - F + G$     (G)
2. $A - B = C + D$     (A)
3. $A \times d = c \times E$     (E)
4. $A = B \times c \times d \times e$     (d)
5. $\dfrac{B}{F} = \dfrac{A}{C}$     (B)
6. $\dfrac{B}{F} = \dfrac{A}{C}$     (F)

### Make the grade

In this section, we have only used the letters A to F. In the next section, we will focus on formulae that have real engineering meaning. In the meantime, to help you understand transposition, have a go at this activity.

# Transposition of engineering formulae

Hopefully you will now be able to rearrange simple formulae using simple letters. The good news is that the basics do not change when we look at real engineering formula – in fact, the only real difference in this section is that the formulae will have genuine engineering meaning.

The formulae used here are explained in more detail in Unit 4 (Applied electrical and mechanical science for engineering).

Let's start with a simple example.

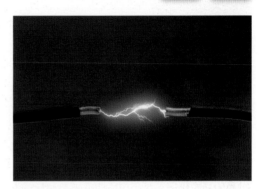

**Figure 3.5**

## Worked Example

**Example 1:**

$$V = I \times R \text{ (also: } P = VI \text{ or } m = \rho V)$$

$V = IR$ is arguably the best-known of engineering formulae – it represents Ohm's law and is used to calculate voltage (V) from the multiplication of current (I) and electrical resistance (R). Transposition would be essential if we knew the voltage and current but wanted to calculate resistance.

$$V = IR$$

To make R the subject, we would need to divide both sides by I and simplify. The formula becomes:

$$\frac{V}{I} = R$$

(The same technique would apply if we knew the voltage and resistance but intended to work out the current.)

$$V = I \times R$$

To make I the subject, we would need to divide both sides by R and simplify:

$$\frac{V}{R} = I$$

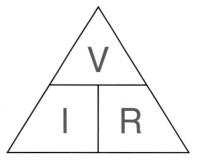

**Figure 3.6** Voltage, current, resistance formula triangle

## Team Talk

Aisha: **'Is there any way you can check the transposition in Example 1 is correct?'**

Steve: **'The use of a triangle is a very popular method of checking when you have a formula such as V = I × R.' (Figure 3.6)**

**'The value at the top (in this case, V) will always be above the division bar if used in a formula. The values at the bottom (in this case, I and R) will always be under the division bar. If you are trying to work out the relationship, simply put your finger over the one you are making the subject and go from there. If they are on the same level, then you multiply them together, so V = I × R. The other possibilities both involve division: I = $\frac{V}{R}$ and R = $\frac{V}{I}$.'**

Steve: **'You need to be careful, though – this only works when the formula has two variables multiplied together to make the third!'**

Aisha: **'Yeah, it could be used in exactly the same way for P = VI or m = ρV.'**

## Worked Example

**Example 2:**

$$P = I^2R, \qquad \text{also } v = \sqrt{2gh}, \quad I = \sqrt{\frac{P}{R}}$$

To take another step forward, we will now consider how to deal with formulae that involve a square and square root of a variable.

If we knew the power (P) of a circuit and also knew the resistance (R), then there may come a need to calculate the current (I). In this case, we would take the basic formula, $P = I^2R$, making I the subject, as follows:

$$P = I^2R$$

First, we would need to move R from the RHS to the LHS. We can do this by dividing both sides by R. The formula becomes:

$$\frac{P}{R} = I^2$$

Now we must remove the $^2$ on the RHS by square-rooting both sides. As the $^2$ and $\sqrt{\phantom{x}}$ cancel on the RHS, the formula can be simplified:

$$\sqrt{\frac{P}{R}} = I$$

## Worked Example

**Example 3:**

A similar situation would arise if we knew the volume (V) and gravity (g) of an industrial container but wanted to determine the height. The formula would be as follows:

$$V = \sqrt{2gh}$$

First, we must remove the √ on the RHS by squaring both sides, so the formula becomes:

$V^2 = 2gh$    (the √ and $^2$ cancel out on the RHS, leaving the $^2$ on the LHS)

Now it is simply a case of moving the $^2$ and the g to the LHS. We can do this by dividing both sides by (2 × g):

$$\frac{V^2}{2g} = h$$

## Activity

Using the formula $I = \sqrt{\dfrac{P}{R}}$, transpose to make R the subject.

Tip: Remove the square root to the LHS by squaring both sides, move the R to the LHS (to move it above the division bar) and go from there!

## Worked Example

**Example 4:**

$s = ut + \dfrac{1}{2}at^2$ (also $v^2 = u^2 + 2as$, $X_c = \dfrac{1}{2}\pi fC$)

These final examples are a little more complex. Yet if you follow the basic rules, they should not cause much additional difficulty.

If we knew the acceleration (a), the time (t) and the distance (s), and we wanted to determine the initial velocity (u) of a car, we would need to rearrange the following formula to make u the subject:

**Figure 3.7**

$$s = ut + \frac{1}{2}at^2$$

The first step is to move $[+\frac{1}{2}at^2]$ to the LHS – we can do this by subtraction of $[\frac{1}{2}at^2]$ from both sides:

$$s - \frac{1}{2}sat^2 = ut + \frac{1}{2}at^2 - \frac{1}{2}at^2$$

After the cancellation we have:

$$s - \frac{1}{2}at^2 = ut$$

The next stage is to move the t to the LHS by dividing both sides by t:

$$\frac{s}{t} - \frac{1}{2}at^2 / t = \frac{ut}{t}$$

After the cancellation, we have:

$$\frac{s}{t} - \frac{1}{2}at = u$$

## Activity

1. a) Using the formula $v = u + at$, make a the subject.
   b) If $v = 10$ m/s, $u = 2$ m/s and $t = 4$ seconds, what is a in m/s²?
2. a) Using the formula $P = I^2R$, make R the subject.
   b) If $I = 5$ amps and $P = 150$ watts, what is R?
3. Using the formula $v = \sqrt{2gh}$, make g the subject.

### Make the grade

This activity will help you in achieving the following grading criterion:

**P2**  use algebraic methods to transpose and evaluate simple formulae.

## Activity

1. a) Using the formula $s = ut + \frac{1}{2}at^2$, make a the subject. Hint: start by moving ut to the LHS and go from there!
   b) If $u = 10$ m/s and $s = 30$ metres, calculate a in m/s².
2. a) Using the formula $v^2 = u^2 + 2as$, make s the subject. Hint: start by moving $u^2$ to the LHS and go from there!
   b) If $v = 15$ m/s, $u = 5$ m/s and $a = 3$ m/s², what is s?
3. Using the formula $Xc = \frac{1}{2\pi fC}$, make C the subject. Hint: start by getting C above the division line. From there it is quite straightforward.

### Make the grade

This activity will help you in achieving the following grading criterion:

**M1**  transpose and evaluate complex formulae.

# Combining formulae

This section looks at what to do when you want to transpose a formula that has the same term on both sides.

## Worked Example

**Example 1:**

$$\tfrac{1}{2}mv^2 = mgh$$

This formula has mass (m) on both sides.

Written out in full, this is:

| $\frac{1}{2}$ | × | m | × | $v^2$ | = | m | × | g | × | h |
|---|---|---|---|---|---|---|---|---|---|---|

If we are transposing to make v the subject, then we will rearrange the formula as follows:

First, we need to remove the $\frac{1}{2}$. We can do this by multiplying both sides by 2. This will remove the $\frac{1}{2}$ on the LHS (as $2 \times \frac{1}{2} = 1$) and add a multiplier of 2 to the RHS. The formula becomes:

$$1 \times m \times v^2 = 2mgh \quad \text{(although we don't usually write in the 1)}$$

Or simply:

$$mv^2 = 2mgh$$

The next step is to collect the m term on the RHS. We can do this by dividing both sides by m:

$$\frac{mv^2}{m} = \frac{2mgh}{m}$$

Following simplification, we have:

$$v^2 = 2gh$$

We can now complete the transposition by removing the $^2$ from v – we do this by square-rooting both sides:

$$v = \sqrt{2gh}$$

**Example 2:**

$$A = 2(B + C)$$

This example looks relatively straightforward initially.

Transpose the formula to find C.

The first stage is to remove the brackets. To do this, multiply the subject inside the brackets by the number outside, so the formula becomes:

$$A = 2B + 2C$$

We can now move the 2B to the LHS by subtraction of the same term to both sides. After simplification:

$$A - 2B = 2C$$

We can now move the 2 from the RHS to the LHS by dividing both sides by 2:

$$\frac{A}{2} - \frac{2B}{2} = \frac{C}{2}$$

Following simplification, we have:

$$\frac{A}{2} - B = C$$

## Activity

- **Using the formula below, make V the subject:**
  $$\tfrac{1}{2}QV = \tfrac{1}{2}CV^2$$
  **Hint: start by multiplying both sides by 2 (to remove the $\frac{1}{2}$s), then divide both sides by V and go from there.**
- **Use the combined formula you have just derived to determine the answer to the following problem: if the potential difference (V) in a part of a circuit is 15 volts and the capacitance is $10 \times 10^{-6}$ farads, determine the charge (C) in coulombs.**

### Make the grade

This activity will help you in achieving the following grading criterion:

**D1** transpose and evaluate combined formulae.
Note: the following activity is not designed to be easy and, in some cases, you will need to seek additional help from your teacher or tutor. If you can solve it, then you are clearly working at distinction level!

# Graphs and plots

Sometimes the terms 'graph' and 'chart' are used interchangeably but the two are actually very different.

Charts are generally used to categorize information into bars, segments or area. A number of common charts are shown in Figure 3.8.

A graph (sometimes referred to as a 'plot') has two continuous axes running vertically and horizontally. The x axis is drawn as an arrow that points horizontally to the right; the y axis is drawn as an arrow that points upwards. The graph is used to show the relationship between a value on the x axis and a value on the y axis. The plotted points are usually joined to form a line.

Velocity–time plots are a good example of a graph and are

**Figure 3.8** Common charts          **Figure 3.9** Graph

discussed in more detail in Unit 4 (Applied electrical and mechanical science for engineering).

It is important to note that the axes do not necessarily have to start at zero (x axis = 0, y axis = 0); if working with large numbers, between say $1.00 \times 10^6$ and $5.00 \times 10^6$, then starting at zero would compress the information being represented. Another point to remember is that the plotted line does not necessarily need to be linear (or straight).

> ## 66 Team Talk
>
> Aisha: **'What does linear mean?'**
> Steve: **'It just means straight – a linear line is a straight line!'** 99

Figure 3.12 represents acceleration and velocity performance projections from the Bloodhound SSC Team. The SSC team currently holds the land-speed record; in October 1997, Thrust SSC (driven by RAF pilot Andy Green, who is also a mathematics graduate from Oxford University) achieved the remarkable land speed of 763.035 m.p.h. in the Black Rock Desert, Nevada, USA. The same team will attempt to break its own record when Andy Green tries to achieve over 1,000 m.p.h. from the Bloodhound's EJ200 jet engine.

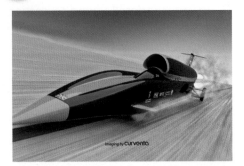

**Figure 3.10** Bloodhound SSC team

**Figure 3.11** Bloodhound SSC team engine

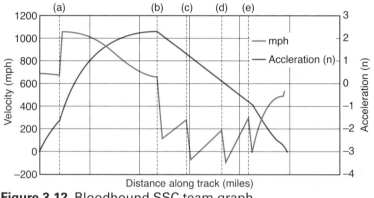

**Figure 3.12** Bloodhound SSC team graph

A few simple rules to follow when drawing a graph include the following:

1. Title your graph.
2. Label the axes.
3. Ensure your measurements and scale are consistent and appropriate for the values you are plotting.

## Activity

Plot the early land-speed records shown in Table 3.7 on a graph. The x axis should plot the year and the y axis should plot the speed. The speed is a very accurate measurement; before plotting the graph, round the speed to the nearest whole number.

| Year | Car | Speed (velocity) in m.p.h | Rounding |
|------|-----|---------------------------|----------|
| 1898 | Jeantaud | 39 | |
| 1899 | Jenatzy | 65.79 | |
| 1902 | Mors | 77.13 | |
| 1904 | Darracq | 104.52 | |
| 1906 | Stanley | 121.57 | |
| 1919 | Packard | 149.87 | |
| 1924 | Sunbeam | 146.16 | |
| 1928 | Triplex | 207.55 | |
| 1939 | Railton | 369.70 | |

**Table 3.7** Land-speed records

## Activity

Using the formula V = IR, complete Table 3.8 by working out the value of resistance.

| Voltage (volts) | Resistance (ohms) | Current |
|---|---|---|
| 10 | 2 | 5 |
| 20 | | 5 |
| 30 | | 5 |
| 40 | | 5 |
| 50 | | 5 |
| 60 | | 5 |
| 70 | | 5 |
| 80 | | 5 |
| 90 | | 5 |

**Table 3.8** Voltage, resistance and current

Now produce a graph plotting the voltage (y axis) and the resistance (x axis).

The gradient of the line $\frac{y}{x}$ at any point will provide you with the current (this is only possible for a (0, 0) graph). When this happens, the voltage and resistance are said to be 'proportional'.

# Learning Outcome 2.  Be able to use mensuration and trigonometry to solve engineering problems

## Simple area and volume

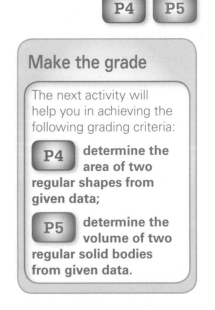

Estimating area and volume is one of the most fundamental aspects of engineering mathematics and it is carried out for a number of reasons. Engineers often need to estimate the material required to make a product or component – examples include the sheet steel to process automotive body work, the volume of polymer used to make electrical casings, the area and volume of pipes used in estimating flow in petrochemical operations, and the volume of ingredients used in food processing or pharmaceutical manufacturing.

Other examples include the design of a new factory layout or the estimation of an area potentially used for engineering office space in a new plant.

You are highly likely to have covered this material at school, so Tables 3.9 and 3.10 should provide useful revision.

### Make the grade

The next activity will help you in achieving the following grading criteria:

**P4** determine the area of two regular shapes from given data;

**P5** determine the volume of two regular solid bodies from given data.

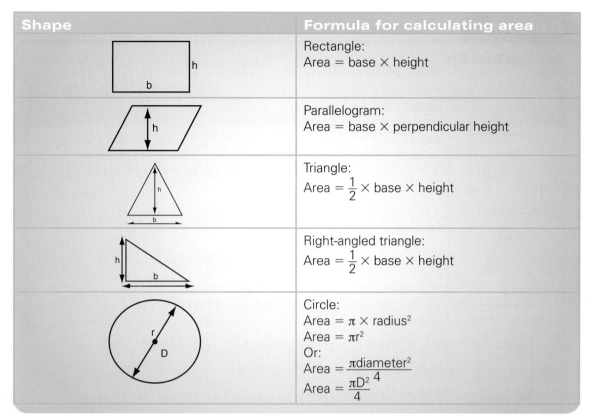

| Shape | Formula for calculating area |
|---|---|
| | Rectangle:<br>Area = base × height |
| | Parallelogram:<br>Area = base × perpendicular height |
| | Triangle:<br>Area = $\frac{1}{2}$ × base × height |
| | Right-angled triangle:<br>Area = $\frac{1}{2}$ × base × height |
| | Circle:<br>Area = π × radius$^2$<br>Area = $\pi r^2$<br>Or:<br>Area = $\frac{\pi \text{diameter}^2}{4}$<br>Area = $\frac{\pi D^2}{4}$ |

**Table 3.9** Calculating basic area

| Shape | Formula for calculating volume |
|---|---|
| | Cube:<br>Volume = base × height × length |
| | Parallelogram:<br>Volume = base × height × length |
| | Triangle:<br>Volume = $\frac{1}{2}$ × base × height × length |
| | Right-angled triangle:<br>Volume = $\frac{1}{2}$ × base × height × length |
| | Cylinder:<br>Volume = $\pi r^2$ × length |

**Table 3.10** Calculating basic volume

## Activity

Complete Table 3.11.

| Diagram | Formula for calculating area/volume | Calculation |
|---|---|---|
| | Cube:<br>Volume = base × height × length | Volume = b × h × l<br>Volume = 2 × 3 × 4<br>Volume = 24 mm$^2$ |
| | | |
| | | |
| | | |
| | | |

**Table 3.11** Formulae for calculating area/volume

# Area and volume of compound shapes

M2    M3

We will now build on our knowledge from the preceding section and focus on compound shapes. These are areas and volumes made up of different types of shapes. A combination of the formulae introduced previously will be required to determine the area (for a two-dimensional shape) and volume (for a three-dimensional shape).

Computer-aided design (CAD) packages are very helpful in achieving this when designing or prototyping new products, although any technician or engineer would need to fully understand the basics. Here, four examples are used to illustrate how to calculate compound areas and volumes.

## Worked Example

### Example 1:

Consider how you would calculate the area of the shape in Figure 3.13.

At first it may look a little tricky, but if you consider that part A is a simple rectangle, part B is also a rectangle and part C is a triangle, then all you have to do to work out the total area is calculate the individual parts and add them all together (i.e. A + B + C).

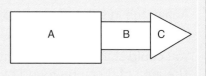

**Figure 3.13**

### Example 2:

Consider how you would calculate the volume of the shape in Figure 3.14.

You should have realised that, although the shape is not specifically covered in Table 3.10, it can still be calculated relatively easily. First, calculate the surface area of face A (by working out the area and then dividing by two). To complete the calculation, you simply need to multiply the area by the length.

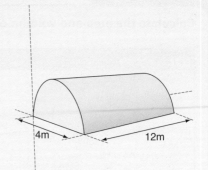

### Example 3:

Slightly more tricky, consider how you would calculate the area of the shape in Figure 3.15.

**Figure 3.14**

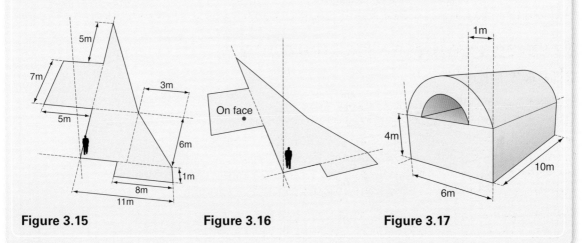

**Figure 3.15**          **Figure 3.16**          **Figure 3.17**

At first, this looks quite difficult, but the easiest way is to simply divide it up into parts you recognise (Figure 3.16).

Now it is simply a case of working out the area of the rectangles and triangles.

**Example 4:**

Consider how you would calculate the volume of the shape in Figure 3.17.

Again, the initial complexity of the compound shape can be simplified by considering each part individually before summing to reach the total volume.

You could start by working out the volume of A (simply, base × height × length). The area of B can be found by taking the area of the smaller semi-circle (b) from the larger semi-circle (c) and then multiplying by the length to get the volume.

## Activity

**Calculate the area and volume of the compound shapes shown in Table 3.12.**

| Diagram | Formulae and working |
|---|---|
| 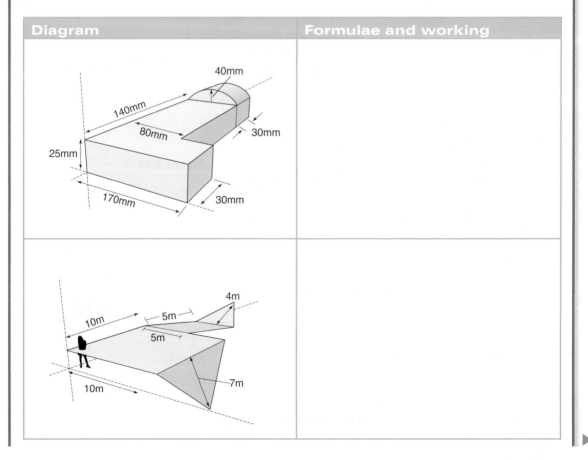 | |

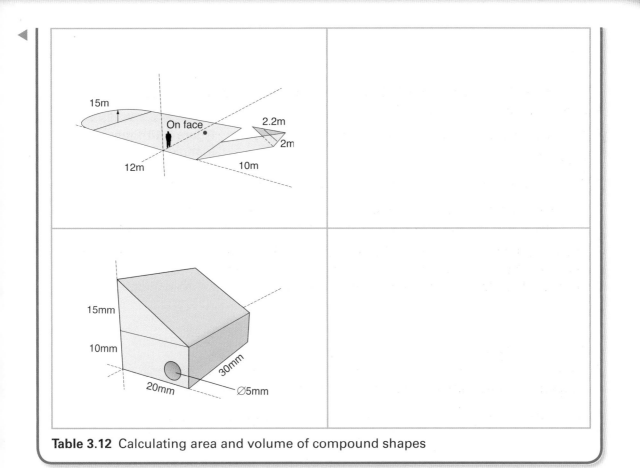

**Table 3.12** Calculating area and volume of compound shapes

## Make the grade

This activity will help you in achieving the following grading criteria:

**M2** identify the data required and determine the area of two compound shapes;

**M3** identify the data required and determine the volume of two compound solid bodies.

# Pythagoras's theorem and basic trigonometry

## Pythagoras's theorem

First of all, a few basics:

- There are four types of triangle: equilateral, isosceles, scalene and right-angled – Pythagoras's theorem and trigonometry are useful only when working with the last of these.

- A right-angled (or plain) triangle is formed when two sides of the three sides come together to form a 90 degree angle. The other two angles vary depending on the lengths of the sides.
- The angles within a triangle all sum to 180 degrees.
- When considering a right-angled triangle, if two sides are known, the third can be calculated using Pythagoras's theorem.

**Figure 3.18** Right-angled triangle

## Worked Example

**Example:**

$$A^2 + B^2 = C^2$$

If A = 8 and B = 6, calculate C.

First, you need to find the square of 8 and 6:

$$A^2 = 8^2 = 8 \times 8 = 64$$
$$B^2 = 6^2 = 6 \times 6 = 36$$

Now add them together:

$$A^2 + B^2 = C^2$$
$$64 + 36 = 100$$

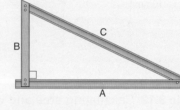

**Figure 3.19**

To complete this calculation, you need to find C.
This is found by applying the square root to $C^2$ to the square root of C, in this case 100.

$$C = \sqrt{100} = 10$$

## Activity

**Fill in Table 3.13.**

| Side A | Side B | Side C |
|--------|--------|--------|
| 3      | 8      |        |
| 125    |        | 12     |
| 1000   | 1000   |        |
| 0.125  | 0.5    |        |

**Table 3.13**

# Trigonometry

Trigonometry has its roots in astronomy and geography, although the historical origins can be traced as far back as the Egyptians, Babylonians and Ancient Greeks. Today, trigonometry (or 'trig') is commonly used by surveyors, civil engineers, military engineers and mechanical engineers for a host of applications, including the design and construction of roads, bridges and structures and in other areas, such as satellite navigation, electronics, seismology, computer graphics and even economics.

In a nutshell, trig is used when you want to calculate an angle or side of a right-angled triangle in the following conditions:

- to find the triangle angles when the lengths of sides are known;
- to find the length of a side if an angle and length of one side are known.

The sides of a right-angled triangle are commonly expressed as:

- hypotenuse: the side opposite the right angle;
- opposite: the side opposite the given angle;
- adjacent: the side next to the given angle.

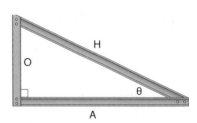

**Figure 3.20**

**Team Talk**

Aisha: **'What is that symbol that has appeared?'**
Steve: **'It's the Greek letter theta (θ).'**
Aisha: **'What is that?'**
Steve: **'It is commonly used in trigonometry to show the angle that is either known or needs to be known.'**

A key point to note is that if the position of θ changes, then the opposite and adjacent sides need to be reclassified. The hypotenuse remains the same, as it is still the side opposite the right angle.

Remember that trigonometry is a mathematical tool to help solve unknown angles and lengths on a triangle, and nothing more. It can be a little puzzling at first, but as long as you follow a couple of basic rules, it is fairly straightforward.

**Figure 3.21**

First of all, it is helpful to remember a simple mnemonic – Soh-Cah-Toa.

**66 Team Talk**

Aisha: **'A what?'**
Steve: **'A mnemonic?' 'It's a learning aid to help remember something important.'**

Soh-Cah-Toa is used to help us remember that, depending on the angle and/or sides we want to find, we will need to use one of the following:

$$\text{Sine}\theta = \frac{\text{opposite side}}{\text{hypotenuse}} \qquad \textbf{(Soh)}$$

$$\text{Cosine}\theta = \frac{\text{adjacent side}}{\text{hypotenuse}} \qquad \textbf{(Cah)}$$

$$\text{Tangent}\theta = \frac{\text{opposite side}}{\text{adjacent side}} \qquad \textbf{(Toa)}$$

**66 Team Talk**

Aisha: **'What are sine, cosine and tangent?'**
Steve: **'Trigonometric functions.'**
Aisha: **'What?'**
Steve: **'It means that they can be used to do a specific thing; the trigonometric ratios they are made from constructing triangles within circles and they are very useful for helping us calculate angles and sides. It sounds complicated but in practice it's just a couple of buttons on your calculator!'**

Before we go any further, it is important that you can work out the sine, cosine and tangent of an angle on your calculator.

Although calculator models and makes vary, any scientific calculator will have the following buttons: [sin] [cos] [tan].

To calculate the sine of a 30 degree angle:
Press the [sin] key and then enter the angle [30].
If that gives you an error on your calculator, just do it the other way round, so:
Enter the angle [30] and press the [sin] key.
Either way you should get the number [0.5].

## Activity

**Using the method above, find the following:**

- [tan] [50.19];
- [sin] [18];
- [cos] [89];
- [sin] [0.16].

To help you with trigonometry, we will work through three examples. If you can follow the examples, by the end you should be able to do trig. If you are still unsure, then ask your teacher or tutor for further instruction.

## Worked Example

**Example 1: how to find the angle if two other angles are known.**

How to solve: this is actually very simple and you do not even need to use trigonometry. All the angles together add up to 180 degrees, so the missing side must simply be 45 degrees [90 + 45 + 45 = 180].

**Example 2: how to find the angle if two or more sides are known.**

How to solve: the first stage is to understand which of the sides are the adjacent, the hypotenuse and the opposite. The 10-metre side is directly opposite the right angle, so this is the hypotenuse. The 8-metre side is opposite θ (the angle we want to know), so this is named the opposite side. The side with an unknown length is, therefore, the adjacent side.

To work out the angle, we must select one of the three parts of Soh-Cah-Toa where we already know two parts – in this example, we know the opposite and hypotenuse sides. Therefore, we will use Soh, which is a mnemonic for:

$$\sin\theta = \frac{\text{opposite side}}{\text{hypotenuse side}}$$

If the opposite side is 8 m and the hypotenuse side is 10 m, then:

$$\sin\theta = \frac{8}{10} = 0.8$$

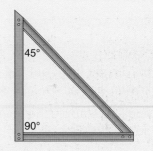

**Figure 3.22**

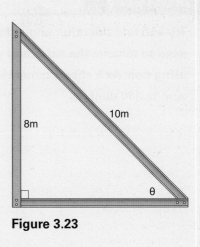

**Figure 3.23**

To complete the example and get θ (the angle) from sinθ (sine of the angle), we must apply the inverse of sin (this is sometimes called the arc-sine). To do this, follow the simple steps below:

First press the [shift] button.
Now press the [sin] button.
Enter your number [0.8].
Finally, press [=] to show the answer in degrees.

Note: on some calculator models, you may have to enter in a slightly different order: [0.8], [shift], [sin], or use a second function button.

The calculator should read 53.13: this is the answer. The angle between the hypotenuse and adjacent sides (originally called θ), we now know is 53.13 degrees. If you have not got this answer check with your teacher – your calculator may be in the wrong mode. It should be in DEG (Degrees) mode.

We could follow the same steps to find any angle from two known sides as long as:

• firstly, we identify which sides are the hypotenuse, the adjacent and the opposite;

• secondly, we select the part of Soh-Cah-Toa that is most appropriate;

• thirdly, we work out the sin, cos or tan of the unknown angle by dividing the two known sides (how you do this depends on if you are using Soh, Cah or Toa);

• finally, we apply the inverse of either sin, cos or tan to find the angle theta (θ) in actual degrees.

## Activity

Try and find the other angle (Figure 3.24). Hint: you will need to rename the sides and you should think about using Cah. As a check, remember the angles should sum to 180 degrees.

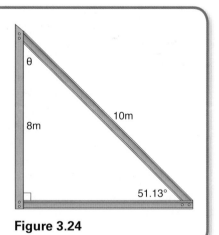

**Figure 3.24**

## Worked Example

**Example 3: how to find the length of a side, if one length of a side and one angle are known.**

How to solve: to start with, we must follow the four steps as above.

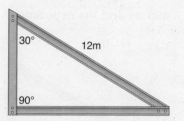

Firstly, identify which side is the hypotenuse, the opposite and the adjacent. The 12-metre side is opposite the right angle, so must be the hypotenuse; the side running horizontally along the bottom must be the opposite side (because it is opposite the 30 degrees); therefore, the vertical side must be the adjacent.

**Figure 3.25**

Secondly, in selecting the part of Soh-Cah-Toa that is most appropriate, we need to consider which side we are trying to determine. Remember, we know the hypotenuse and we know θ (the angle) – we can, therefore, work out the adjacent side by applying Cah or the opposite side by applying Soh.

Thirdly, we use Cah to find the adjacent side (we have selected Cah as we know two of the three values, i.e. cosθ and the hypotenuse, and we want to know the third, i.e. the adjacent side):

$$\cos\theta = \frac{\text{adjacent}}{\text{hypotenuse}}$$

$$\cos 30° = \frac{\text{adjacent}}{12} \text{ (we can work out cos30° by keying it into our calculator)}$$

$$= 0.87$$

Rearranging the formula gives:

$$\text{adjacent} = 0.87 \times 12 = 10.44 \text{ metres}$$

Finally, we use Soh to find the opposite side (we have selected Soh as we know two of the three values, i.e. sinθ and the hypotenuse, and we want to know the third, i.e. the opposite side):

$$\sin\theta = \frac{\text{opposite}}{\text{hypotenuse}}$$

$$\sin 30° = \frac{\text{opposite}}{12} \text{ (we can work out sin30° by keying it into our calculator)}$$

$$= 0.5$$

Rearranging the formula gives:

$$\text{opposite} = 0.5 \times 12 = 6 \text{ metres}$$

## Activity

1. Using Pythagoras's theorem, determine the length of the opposite side in Figure 3.26 if the adjacent side is 15 metres and the hypotenuse is 22 metres.

2. Using trigonometry, determine the angle $\theta$ in Figure 3.27 if the adjacent side is 15 metres and the hypotenuse is 22 metres.

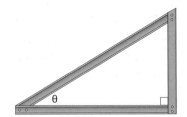

**Figure 3.26**

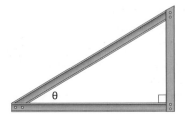

**Figure 3.27**

## Activity

Use trigonometry to determine the following in Figure 3.28:

- the height of the trapezoid;
- the lengths of the sloping sides.

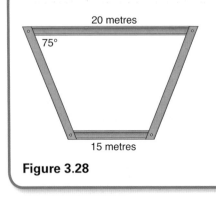

**Figure 3.28**

## Make the grade

The activity above will help you in achieving the following grading criterion:

**P6** solve right-angled triangles for angles and lengths of sides using basic Pythagoras's theorem, sine, cosine and tangent functions.

## Make the grade

This next activity will help you in achieving the following grading criterion:

**M4** use trigonometry to solve complex shapes.

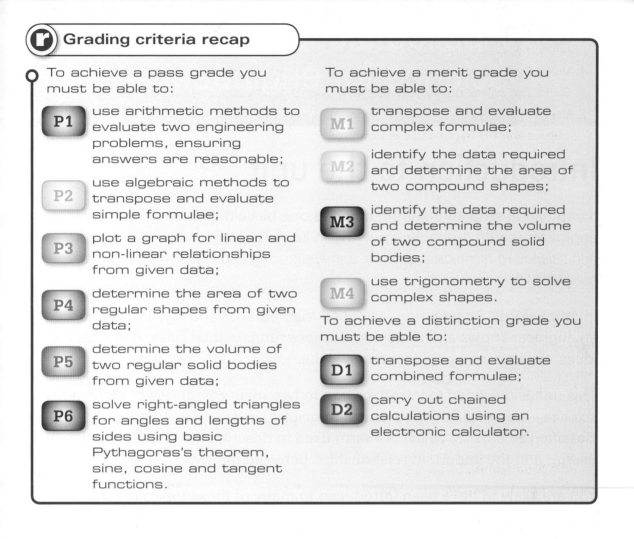

**Grading criteria recap**

To achieve a pass grade you must be able to:

**P1** use arithmetic methods to evaluate two engineering problems, ensuring answers are reasonable;

**P2** use algebraic methods to transpose and evaluate simple formulae;

**P3** plot a graph for linear and non-linear relationships from given data;

**P4** determine the area of two regular shapes from given data;

**P5** determine the volume of two regular solid bodies from given data;

**P6** solve right-angled triangles for angles and lengths of sides using basic Pythagoras's theorem, sine, cosine and tangent functions.

To achieve a merit grade you must be able to:

**M1** transpose and evaluate complex formulae;

**M2** identify the data required and determine the area of two compound shapes;

**M3** identify the data required and determine the volume of two compound solid bodies;

**M4** use trigonometry to solve complex shapes.

To achieve a distinction grade you must be able to:

**D1** transpose and evaluate combined formulae;

**D2** carry out chained calculations using an electronic calculator.

# Unit 4
## Applied electrical and mechanic science for engineering

# Introduction to the unit

If you had to strip engineering down to one basic definition, then many engineers would probably agree that it is: *the application of mathematics and science to technical problems, and suggesting and implementing suitable solutions.*

In other words, engineering is applied mathematics and science. The more an engineer knows and understands these fundamental rules, methods and ideas, the better an engineer they will be.

This unit is designed to introduce you to two main categories of engineering science – electrical and mechanical principles. In essence, the topics can be classified as *physics*, which is a term used to describe the study of matter, and *energy*, and the important relationships between the two.

You are likely to have been introduced to many of these topics at school through your GCSE work in maths or science – this unit will take these concepts and apply them to the engineering world.

### Learning Outcomes

By the end of this unit you should:

- be able to define and apply concepts and principles relating to mechanical science;
- be able to define and apply concepts and principles relating to electrical science.

# Grading criteria

| To achieve a pass grade you must be able to: | To achieve a merit grade you must be able to: | To achieve a distinction grade you must be able to: |
|---|---|---|
| **P1** define parameters of direct-current electricity and magnetic fields | **M1** determine the force on a current-carrying conductor situated in a magnetic field from given data | **D1** explain the construction, function and use of an electro-magnetic coil |
| **P2** determine total resistance, potential difference and current in series and parallel DC circuits from given data | **M2** describe the conditions required for the static equilibrium of a body | **D2** determine the work done and the power dissipated in moving a body of given mass along a horizontal surface at a uniform velocity, given the value of the coefficient of kinetic friction between the contact surfaces |
| **P3** define parameters of static and dynamic mechanical systems | | |
| **P4** determine the resultant and equilibrant of a system of concurrent coplanar forces from given data | | |
| **P5** determine the uniform acceleration/retardation of a body from given data | | |
| **P6** determine the pressure at depth in a fluid from given data | | |

# Units, multipliers and prefixes

When you are learning any subject for the first time, there are always a couple of basics you need to learn first. When learning to drive, you need to understand how to use the clutch and when learning a foreign language, you may start with basic numbers. That is what we are going to do here: start with a few numbers.

Engineers often use very large or very small numbers to express quantities. To make things easier, they may use a kind of shorthand in discussion and when producing technical documentation.

**Figure 4.1**

The chances are that you will have used this language yourself without even knowing it: 'My iPod has 16 gigabytes', or 'My phone is only 6 millimetres thick.' To you, that refers to the memory capacity of your media player and the depth of your phone; to engineers, it provides three basic types of information, which are summarised in Table 4.1.

|              | Unit    | Multiplier          | Prefix |
| ------------ | ------- | ------------------- | ------ |
| **iPod**         | Bytes   | $16 \times 10^9$    | giga   |
| **Mobile phone** | Metres  | $6 \times 10^{-3}$  | milli  |

**Table 4.1** Units, multipliers and prefixes

Do not worry if that does not make sense yet – it simply shows how engineering science underpins your everyday language without you even noticing.

Units are used to standardise measurement. This means that, if an engineer in Japan and a designer in the United States are collaborating on a project, then they would know exactly what each other is talking about – for example, if they said, '3 amps of current'.

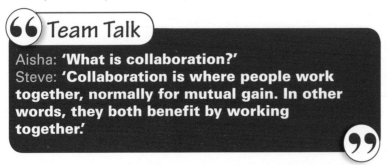

**66 Team Talk**

Aisha: **'What is collaboration?'**
Steve: **'Collaboration is where people work together, normally for mutual gain. In other words, they both benefit by working together.'**

The International System of Units or SI system (from the French *le Système International d'Unités*) is widely adopted to define the use and application of units – again, you may have used this system in everyday language if you have ever referred to the 'metric system'.

The system is based on seven 'base units' from which a host of 'derived units' have been established.

The base units are shown in Table 4.2.

| Unit | Symbol | Use |
| --- | --- | --- |
| kilogram | kg | mass |
| metre | m | length |
| second | s | time |
| ampere | A | electric current |
| kelvin | K | thermodynamic temperature |
| candela | cd | luminous intensity |
| mole | mol | amount of substance |

**Table 4.2** The base units of the International System of Units

The derived units include the joule for energy, the watt for power and the ohm for electrical resistance. These units can be derived (or put together) from the seven base units. The joule, for example, is a product of length, mass and time.

In many cases, the derived units are named after the person who was highly influential in first understanding the science associated with them. The following activity will help you examine five of these historical engineers and scientists in more detail.

The second part of the SI system is concerned with the use of prefix and multiplier. A prefix is a term added to a quantity to give it either greater or lower magnitude (size).

You may say, 'I am going to drive 20 kilometres' – this is actually 20,000 metres. 'kilo' is the prefix for 1,000. I am sure you will agree, it is easier to say (and write) '20 kilometres' than '20,000 metres'.

Another example is '16 gigabytes'. The alternative is '16 thousand million bytes', or '16 billion bytes' – either way, '16 gig' is much easier.

The list of prefixes shown in Table 4.4 is not exhaustive, as some of the quantities have limited application for engineers who generally work in multiples of three (i.e. $10^{-3}$, $10^{-6}$, $10^3$, $10^6$, $10^9$). Other prefixes do not have much

# Activity

Table 4.3 lists a number of units that were named after the engineers and scientists who were highly influential in developing the science associated with them. Find out what quantity each unit measures and why the people associated with them were so important.

| Unit | Measures | Person |
|---|---|---|
| newton | force | Sir Isaac Newton: mathematician and physicist who discovered gravity, calculus and laws of motion |
| farad | | |
| ohm | | |
| weber | | |
| watt | | |

**Table 4.3** Influential engineers and scientists

use as they are either very large or very small for any real application. For example, $10^{100}$ is written out in full as:

10,000,000,000,000,000,000,000,000,000,000,000,0
00,000,000,000,000,000,000,000,000,000,000,000,0
00,000,000,000,000,000,000,000. This has practically no everyday use, yet you will use the prefix regularly, when referring to an internet search engine. It is a 'google'.

| Number | Factor | Prefix | Example |
|---|---|---|---|
| 1,000,000,000 | $10^9$ | giga (G) | My iPod has 120 gigabytes of memory. |
| 1,000,000 | $10^6$ | mega (M) | The resistor has a mega-ohm of resistance. |
| 1,000 | $10^3$ | kilo (k) | I have lost 1 kilogram due to my diet. |
| 10 | $10^1$ | deca or deka (da) | A deca is simply a multiple of 10. For example, $2 \times 10^1$ is 20. |
| 1 | $10^0$ | | This is simply the number 1. |
| 0.01 | $10^{-2}$ | centi (c) | My little brother is 80 centimetres tall. |
| 0.001 | $10^{-3}$ | milli (m) | My phone is 6 millimetres thick. |
| 0.000001 | $10^{-6}$ | micro ($\mu$) | The layer of paint on the aircraft is only 30 microns thick. |
| 0.000000001 | $10^{-9}$ | nano (n) | Scientists are leading the way in nano-medicine and nano-robotics. |
| 0.000000000001 | $10^{-12}$ | pico (p) | A very small capacitor, perhaps the smallest in everyday use, is about 1 picofarad. |

**Table 4.4**  Prefixes

## Activity

Have a go at completing Table 4.5 below by converting between the prefix, multiplier and base numbers.

Your teacher should explain the best way of moving between the quantities, which usually involves moving numbers from the right of the decimal point to the left (if the base number is larger than the prefix number) or from the left of the decimal point to the right (if the base number is smaller than the prefix number).

For example, consider the first quantity in the table, 10 kilometres.
If written in full with a decimal point, it would be 10.0 kilometres.
As kilo means $10^3$, the decimal point should move three times to the right, each movement producing an extra zero.

So:
$10.0 \times 10^3$ (or $10 \times 10 \times 10$) becomes:
$100.0 \times 10^2$ (or $10 \times 10$) when one zero is moved to the left of the decimal point, which becomes:
$1000.0 \times 10^1$ (or simply 10) when another zero is moved to the left of the decimal point, which becomes:
10000.0 when a third zero is moved to the left of the decimal point, i.e. the actual number written in full.

| With prefix | With multiplier | Base number with unit |
|---|---|---|
| 10 kilometres | $10 \times 10^3$ | 10,000 metres |
| 123 millimetres | | 0.123 metres |
| 10 picofarad | | |
| 32 gigabytes | $32 \times 10^9$ | |
| 12 mega-ohms | | |
| 68 micrometers (microns) | | 0.000068 |

**Table 4.5** Prefixes, multipliers and base numbers

**Top tip: the 'ENG' button on your calculator can be useful when checking your answers.**

## Make the grade

There are a number of websites that help you convert between quantities. For example, www.convert-me.com or www.onlineconversion.com. **P/M/D** – although no specific grading criteria are attributed to this section, the content is fundamental to all topics. As such, the previous content should prove useful as underpinning content for all pass, merit and distinction criteria.

# Learning Outcome 2. Be able to define and apply concepts and principles relating to mechanical science

## The 'First' guide to mechanical science

P3  P6

When applied to the BTEC First in Engineering, mechanical science covers topics that govern everything from the solar system, to building mega-structures, to the science behind computer motorsport racing games, to why your ears go weird on an aeroplane, to why, if you measure your weight and say you are 63.5 kg (approximately 10 stone), you would be wrong (even if you were actually 63.5 kg).

Mechanical science covers a range of topics classed as 'statics', 'motion' and the 'behaviour of fluids'. But first it is worthwhile taking a little time to introduce (or brush up on) a few terms that are essential in understanding science. It is also important to note that, in reality, many of these fundamental laws can act and work concurrently on objects. Classical science generally separates and introduces them in ideal conditions – this is purely to help understanding.

## Mass

Mass is a term that often gets mixed up with weight, but technically the two are not the same (however, they are related).

Put simply, mass is a measure of how much substance (or matter) you have. If you had a mass of 63.5 kg, then you would be made up of 63.5 kg of substances such as hydrogen, oxygen, nitrogen, carbon, calcium, phosphorus and so on – all the elements that make you 'you'.

**Figure 4.2** The Moon

Other examples include a bag of regular sugar, which has a mass of 1 kg, and the average car, which is about 1,200 kg (or 1.2 metric tonnes because 1,000 kg = 1 tonne).

Mass (unlike weight) is not affected by gravity – Neil Armstrong's mass on the Moon was the same as when on Earth, even though his weight was not. The reason? Gravity.

> **Key words**
>
> **Mass** is a measure of how much substance (or matter) a material has.
>
> **Gravity** is a force of attraction between any objects that have mass.

## Gravitational acceleration

Gravitational acceleration or gravity is, without doubt, one of the most important things in the universe. Gravity is what keeps the planets in orbit around the Sun; it is what keeps the Moon in orbit around the Earth; it makes the sea tides come in and go out; and it is what keeps us firmly fixed on the ground.

**Figure 4.3** The solar system

In simple terms, gravity is a force of attraction between any objects that have mass. The larger the mass, the larger the force of attraction. If no other forces are present, then this force causes the two objects to accelerate towards each other.

The acceleration due to the gravitational pull of the Earth is 9.80665 m/s², often rounded to 9.81 m/s². This essentially means that after one second, an object falling towards Earth would be travelling at 9.81 metres per second; after two seconds, it would be travelling at 9.81 × 2 = 19.62 m/s; after three seconds, it would be travelling at 9.81 × 3 = 29.43 m/s, and so on, increasing by 9.81 metres per second every second until the object reaches terminal velocity. Terminal velocity is the point at which the speed does not change because the opposing force of drag (i.e. upward force caused by the air resistance) equals the weight of the object.

The Sun's substantial gravity (it has a mass of $1.98892 \times 10^{30}$ kilograms) holds the Earth and the seven other solar-system planets in orbit around it. (Pluto was reclassified a few years ago to 'dwarf planet', along with an asteroid, Ceres.)

The concept of gravity was first proposed by Sir Isaac Newton and published in 1687 in his greatest work and catchily titled *Philosophiae Naturalis Principia Mathematica*. (Translated from Latin, it means 'mathematical principles of natural philosophy'.) Some time before this (in the 1660s), Newton started thinking more deeply about gravity after famously witnessing an apple fall from a tree (although not on his head as some newspaper cartoons at the time suggested).

**Figure 4.4** Sir Isaac Newton

### 66 Team Talk

Aisha: **'Who was Sir Isaac Newton?'**
Steve: **'Newton was a great genius. He is widely attributed as the father of modern-day physics and mathematics, having discovered the constituents of white light (the spectrum) and calculus. He also lectured on light at the University of Cambridge, acted as a Member of Parliament and was Warden of the Royal Mint.'** 99

## Weight

Weight is another name for the attractive gravitational force between an object and another, very large object, such as Earth, a planet or a moon.

It is calculated by multiplying the mass (in kg) by the gravitational acceleration for the large object (in m/s$^2$). The unit of weight and force is the newton (N), named after Sir Isaac.

As the Moon has much less mass than the Earth (about one-sixth), the gravitational attraction between an object

### Key word

**Weight** is the attractive gravitational force between an object and another, very large, object.

and the Moon is less than the gravitational attraction between the same object and the Earth. So, even though the mass of an object would be the same on both the Moon and the Earth, the weight would be different because:

Weight (newtons) = mass of object (kg) × gravitational acceleration (m/s²)

$$\text{Weight} = m \times g$$

Therefore, our 1 kg bag of sugar has a weight of 9.81 newtons (1 kg × 9.81 m/s²) on the Earth. But on the Moon, where the acceleration due to gravity is around 1.63 m/s², our 1 kg bag of sugar has a weight of only 1.63 newtons (1 kg × 1.63 m/s²) – approximately one-sixth of the weight on the Earth. Similarly, our average car has a weight of 11,772 newtons (1200 kg × 9.81 m/s²) on the Earth but 1,956 newtons (1200 kg × 1.63 m/s²) on the Moon.

## Worked Example

**If a pilot has a mass of 60 kg on Earth, what would his weight be?**

Solution:

As weight (newtons) = mass (kg) × gravitational acceleration (m/s²)

Then:

weight = 60 kg × 9.81 m/s² = 588.6 newtons

**If the same pilot were to fly a mission to the Moon, what would his weight be?**

Solution:

As weight (newtons) = mass (kg) × gravitational acceleration (m/s²)

Then:

weight = 60 kg × 1.62 m/s² = 97.2 newtons

## Activity

**Calculate your own weight on Earth and on the Moon**

**Your mass:** _____

**Weight (newtons) = mass (kg) × gravitational acceleration (m/s²)**

**Weight on Earth:** _____

**Weight on the Moon:** _____

# Force

As we now know, weight is actually a force. So force, like weight, is measured in newtons. A force is essentially a push or a pull caused when two or more objects interact with each other. When you kick a football, pass a rugby ball or hit a golf ball, you apply a force.

Force is often described in terms of static force (for example, a stationary object on a table or in the design and construction of a bridge) and dynamic force (for example, a moving object such as a person running, or a car or aeroplane in motion).

Force can be calculated using the same formula used for weight, the only difference being, for general use, we replace gravitational acceleration (g) with acceleration (a). The formula below is called Newton's second law of motion and applies to all forces. The calculation of weight is a special case of Newton's second law:

force (newtons) = mass (kg) × acceleration (m/s$^s$)

## Key words

A **force** is a push or a pull caused when two or more objects interact with each other.

The **density** of a body is defined as the amount of mass in a specific volume.

---

## Worked Example

**If a player kicks a football so that it accelerates at 9.5m/s$^2$ and the ball has a mass of 400 grams, what force did he kick the ball with?**

Solution:

force (newtons) = mass (kg) × acceleration (m/s$^2$)

force = 0.4 kg × 9.5 m/s$^2$ = 3.8 newtons

**Figure 4.5**

---

# Density

If you stuffed a box measuring 1 cubic metre with crumpled pieces of paper and then filled another box (the same size) with brand new packs of paper, then the latter (with new paper) would be denser. Put simply, the box would contain more paper (or mass) in the same space (or volume).

Another example would be if you filled two same-sized boxes, one with old bits of mild steel and the other with bits of packing foam – in this case, the former (with mild steel) would be denser.

The density ($\rho$, pronounced 'rho') of a body is defined as the amount of mass (in kilograms) in a specific volume (in m³). It forms a very simple calculation:

$$\text{density} = \frac{\text{mass}}{\text{volume}}$$

$$\text{Or } \rho = \frac{m}{V}$$

The standard unit for density is, therefore, derived from the kilogram divided by metres cubed: kg/m³.

## 66 Team Talk

Aisha: **'I don't get how the units are formed.'**
Steve: **'Remember the standard units introduced earlier on?'**
Aisha: **'Yeah.'**
Steve: **'Well, it means that they can be used to produce new units.'**
Aisha: **'I still don't get it.'**
Steve: **'Well, the units for density are derived by dividing mass by volume. Mass is measured in kg and volume is measured in m³. Therefore, the units of density are kg/m³.'**

99

## Relative density

The relative density of an object is a measure of its density by comparing it with the density of pure water measured at 4°C. The density of water under these conditions is 1,000 kg/m³.

A density of 1,000 kg/m³ essentially means that 1 m³ of water (i.e. 1 m × 1 m × 1 m) has a mass of 1,000 kg; in other words, 1 metric tonne. This actually means that a big bath of water could have the same mass (and weight) as a small family car.

Aluminium has a density of 2,700 kg/m³. Therefore, it has a relative density of $\frac{2700}{1000}$ = 2.7.

Note: relative density has no units. This is because the units in the top of the formula (kg/m³) are the same as those units on the bottom (they cancel each other out).

### Key word

The **relative density** of an object is a measure of its density by comparing it with the density of pure water measured at 4°C.

**Figure 4.6** Many people are surprised by how much water weighs

## Pressure

Pressure is defined as the force acting on a unit area and is calculated as:

$$\text{Pressure} = \frac{\text{force}}{\text{area}}$$

There are two types of pressure: absolute pressure and gauge pressure.

Absolute pressure is derived by comparison with a vacuum (where no pressure exists) and is a true reading of pressure from a scale starting at zero – absolutely no pressure. Gauge pressure is derived by comparison with atmospheric conditions due to the pressure of the air around us. A car tyre pumped up to 2 bar is a reading of gauge pressure; a punctured tyre may have zero gauge pressure but would still have some absolute pressure inside it consistent with the surrounding atmospheric air conditions.

absolute pressure = gauge pressure + atmospheric pressure

Pressure is derived from the quantities that we have already introduced: density (mass and volume), gravity and also height (or depth).

A submarine faces considerable pressure when submerged; this is due to the weight and density of the sea water. The deeper the submarine dives, the greater the mass (and, therefore, the weight) of sea water above it. If a submarine were to (hypothetically) submerge in crude oil, it would be under less pressure because crude oil has a relative density of roughly 0.9, which is less than that of water (which has a relative density of 1.0).

The quantities are linked by the following formula:

pressure (in a static fluid) = density × gravity × height

$P = \rho \ (kg/m^3) \times g \ (9.81 \ m/s^2) \times h \ (metres)$

Pressure is measured in the pascal (Pa) where $1 \ Pa = 1 \ N/m^2$

**Key word**

**Pressure** is defined as the force acting on a unit area.

## Worked Example

A diving cylinder is dropped into a pool of sea water with a depth of 1.75 metres. If the density of the water is 1,026 kg/m³, what is the pressure?

Solution:

Pressure (in a static fluid)  = density × gravity
  × height

= 1026 × 9.81 × 1.75  **Figure 4.7**

= 17613.855 Pa

= 17.61 KPa (kilopascal)

Other well-known units of pressure are PSI (pounds per square inch) and bar.

## Activity

Complete Table 4.6 by filling in the blanks.

|  | N/m² | Pa | PSI | Bar |
|---|---|---|---|---|
| **Example** | 1 | 1 | 0.000145 | 0.00001 |
| **Car tyre** | 200,000 | | 29.01 | |
| **Diver air cylinder** | | | 3,623 | 250 |
| **Pressure washer** | 10,342,135 | | 1,500 | |

**Table 4.6**  Units of pressure

## Activity

Give a simple description of the following terms. Make sure you include the standard unit. An example is given for the first term.

1. **Mass**

   *A measure of the amount of substance (or matter) a material possesses. Mass is measured in the kilogram.*

2. **Gravitational acceleration**

3. **Weight**

4. **Force**

5. **Density**

### Make the grade

This activity will help you in achieving the following grading criterion:

**P3**  define parameters of static and dynamic mechanical systems.

## Activity

A stirring device (agitator) is sunk into a vessel of crude oil to a depth of 3.26 metres. If the specific density of crude oil is 0.8, calculate the pressure.

# Static-force systems

In this section, we will discuss static-force systems. This is a term used in mechanical science to describe force systems and structures that are stationary. In the next section, we will describe force systems that are dynamic (this means they are moving).

Figures 4.8–4.11 show examples of different types of force systems.

**Figures 4.8 and 4.9** Static systems

**Figures 4.10 and 4.11** Dynamic systems

On close scrutiny, static-force systems experience movement that may, in some cases, be too small to notice. The Golden Gate Bridge and Burj Al Arab in Dubai (Figures 4.8 and 4.9) experience movement due to climate and environmental conditions, the weight of traffic and pedestrians, and any tremors or seismic activity in the ground. To clarify what makes a static-force system: it is when, in normal conditions, the forces in a system balance.

# Newton's laws of motion

Sir Isaac Newton formed three fundamental laws of motion that underpin much of the physical science associated with statics and dynamics.

**Newton's laws of motion:**

1. Every object will remain in a state of motion or rest unless an external force is applied.
2. The relationship between an object's force, mass and acceleration is $F = m \times a$.
3. Every action has an equal and opposite reaction.

We introduced Newton's second law earlier in the unit and we will discuss the first law later. Let us now look at the third (and arguably most famous) law. Here is a simple example.

If you were to stand on a firm bridge made of hardwood, as long as the bridge were stable enough, it would hold your weight (force acting down). In this case, a force due to your weight acts downwards on the bridge; the reaction force from the bridge produces an upward force acting on you that balances with your weight. Because the two forces are equal in size but opposite in direction, there is no resultant force acting on the system (you and the bridge), which is stable. If you were to put more force on the bridge (by adding the weight of another person), then the bridge may or may not be able to produce a reaction force big enough to support the weight of the two people. If it held, the bridge would produce an equal reaction force to remain a stable structure. If the bridge were unable to produce a reaction force to equal the weight of the two people, then it would become unstable. This would produce a resultant force downwards and the bridge would fail.

# Static equilibrium

Engineers need to consider this balance of force when designing any stable structure – everything from a table lamp to a mountain bike. If they did not, it would possibly result in catastrophic failure. In a force system that balances, the system is said to be in equilibrium (Figure 4.12).

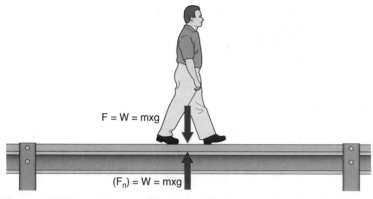

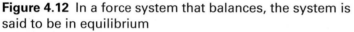

$F = W = mxg$

$(F_n) = W = mxg$

**Figure 4.12** In a force system that balances, the system is said to be in equilibrium

A simple example is a game of tug-o-war. If two people of the same strength were to pull on each opposing end of the rope, then the system would balance (ignoring the contestants' levels of skill and technique). The knot in the rope would remain in the middle of the play area. The system would be in a stable condition – it would be in static equilibrium – because the knot in the rope would not be moving (Figure 4.13).

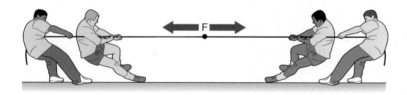

F

**Figure 4.13** The system is in a stable condition

If you were to add a third player to one side, then the knot in the rope would move from the middle towards the team with three people. The system would no longer be stable and would not be in equilibrium. However, for each person, there would still be two forces acting – the action (person pulling on the rope) and the reaction (rope pulling on the person).

To balance the system, returning it to a stable condition and equilibrium, another contestant would need to be added to balance the teams. The forces would then balance, providing a stable structure and one of static equilibrium (Figure 4.14).

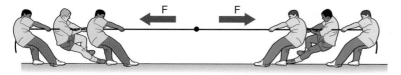

**Figure 4.14** If the same number of contestants were added to each team, the system would again be in a stable condition

What would happen in Figure 4.15?

To solve this relatively complex situation, we can use vector representation. It sounds complex, but it is simply the use of diagrams to illustrate the static-force system. As long as a handful of rules are followed carefully, the method is fairly straightforward.

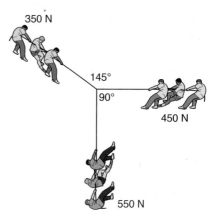

**Figure 4.15**

## 66 Team Talk

Aisha: **'What's a vector?'**
Steve: **'A vector is a quantity that has both size (magnitude) and direction. In other words, if you said you were travelling at 30 km/hr south, then the 30 km/hr provides you with the size of the quantity and 'south' provides you with the direction.'**
Aisha: **'What happens if I forgot to add "south"?'**
Steve: **'Then you wouldn't have any direction. This type of quantity is called a scalar – it just has size. For example, 30 m/s, 10 kg or 144 newtons.'** 99

# Space diagrams, vectors and the polygon of forces

To determine if a system is in equilibrium and therefore stable, you must first have a 'space diagram'. Space, in this instance, refers to the components of the system in the place of interest. It could be a part of a mountain-bike frame or a section of a bridge. In our example above, it is a game of tug-o-war.

If we were to turn Figure 4.15 into a space diagram, we would need to know exactly how much force was being pulled on the rope. We could measure this using the appropriate gauge (or measuring apparatus), but here we will assume that each contestant is pulling with 50 newtons of force. We also need to assume that they are all pulling on the same plane (in this case, the plane is a flat surface parallel to the floor). We could then produce the space diagram, which would look something like Figure 4.16.

In this example, the space diagram is essentially the view looking down on the two teams. To produce the diagram, we have drawn the vector arrow in the same direction as the team is pulling the rope. It is essential to draw the arrow (vector) to scale. For example, 10 newtons equals a line drawn 1 cm long, 50 newtons equals a 5 cm line, 100 newtons equals a 10 cm line, etc. You can use any scale you wish, as long as you are consistent.

To find out if the system is stable and therefore in equilibrium, we need to redraw the space diagram to produce the vector diagram. It is essential that the vectors are drawn accurately in terms of length (representing the size of the force) and direction. To do this, we place each arrow 'nose to tail' with each other. You can start with any arrow, although it tends to be easiest to start with a line acting exactly horizontally or vertically if possible.

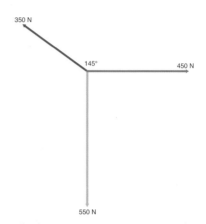

**Figure 4.16** Space diagram

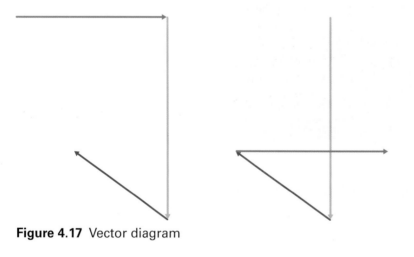

**Figure 4.17** Vector diagram

In this example (Figure 4.17), the arrows do not form a closed system – this means the system is not in equilibrium and is therefore unstable. In this instance, a resultant force would be produced (Figure 4.18). The resultant force can be predicted using this graphical representation – it is the gap between the arrows drawn in the opposite direction (the vector in orange). You will note that, in both examples, the vector has the same-sized force acting in the same direction.

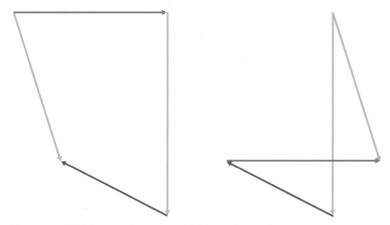

**Figure 4.18** Vector diagram with resultant force

To explain what this means for our game of tug-o-war, we need to return to our space diagram. If the two teams were to pull with the force and direction as shown on the original space diagram, then the knot in the middle would move in the direction and with the force shown by the orange vector (Figure 4.19).

To balance the game and therefore the system, we need to add another force called an equilibrant. This is shown in Figure 4.20 as the grey line, which is exactly the same size as the orange line, just in the opposite direction. To determine the size and direction of this equilibrant (equalising force), we measure the angle and the length of the line, converting back from centimetres to newtons (1 cm = 10 newtons).

**Figure 4.19** Amended space diagram

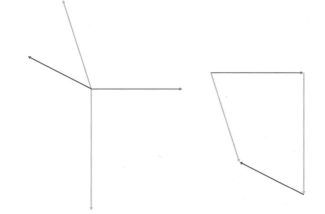

**Figure 4.20** Space diagram with equilibrant and closed-vector diagram

Redrawing the vector diagram, we can see the arrows close, forming a stable structure in equilibrium.

To balance our game of tug-o-war, we would therefore need to add a fourth team, pulling with this amount of force in the exact direction.

This process can be applied to any static-force system that acts on the same plane (coplanar), on the same point (concurrent). This system is described as a polygon of forces system – polygon meaning 'many sides' – and is one of the more complex examples required for the First Diploma.

## The triangle and parallelogram laws

The triangle and parallelogram laws are used to determine the combined magnitude and direction of two forces in a system. Now we have introduced the more complex 'polygon of forces', these two should seem much more straightforward.

The triangle law is essentially what we have discussed already, with the arrows placed 'head to toe' to produce a resultant vector (the orange arrow in Figure 4.21).

**Figure 4.21** Triangle law

The parallelogram law is the diagonal vector that would be produced if the two forces were used to construct a parallelogram (Figure 4.22).

Note: using the triangle and parallelogram laws, the orange resultant force is the same size and in the same direction.

**Figure 4.22** Parallelogram law

## Activity

**Using either the parallelogram or triangle law of vector addition, determine the resultant forces in Figure 4.23.**

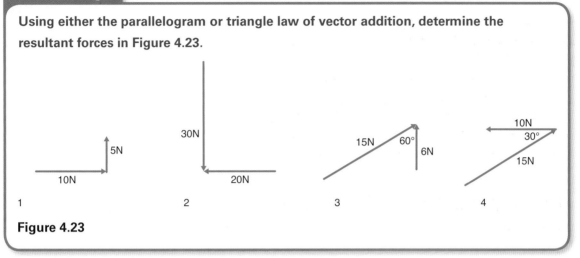

**Figure 4.23**

## Moments of a force

It is relatively straightforward to understand a moment of force. It is the science behind tools such as the Allen key, spanner or socket set.

To illustrate: if you wanted to release a stubborn nut or bolt, which would you prefer – a short or a long spanner? Common sense will tell you that it would be easier with a large spanner because the length of the spanner acts as a multiplier. The longer the spanner, the more force it produces.

A moment is calculated by:

moment = force (in newtons) × perpendicular distance
(in metres)

$$M = F \times s$$

The units are newtons × metres, abbreviated to the newton.metre or Nm.

A moment is a turning force, sometimes referred to as torque – you may have heard of torque when hearing discussions about car engine capacity, because it is produced as a result of the crankshaft and the pistons. Basically, it is used to determine potential performance of the vehicle: the higher the rate of work from the pistons, the quicker the crankshaft turns and the more torque is produced.

## Worked Example

**If a force of 30 newtons is applied to an adjustable spanner with a length of 160 millimetres, calculate the moment of force.**

Solution:

$$M = F \times s$$

$M = 30 \times 0.16$ (note the 160 millimetres has been converted into metres)

$$M = 4.8 \text{ Nm}$$

## Activity

**If a turning force of 6.7 Nm has been produced by a force of 40 newtons, what is the length of the spanner?**

## Activity

Construct a vector diagram based on the space diagram in Figure 4.24 using the polygon of force rule. Is the system in equilibrium? If not, determine the resultant and equilibrant forces.

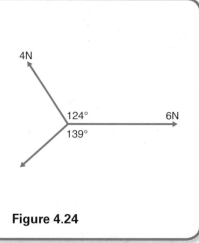

**Figure 4.24**

### Make the grade

This activity will help you in achieving the following grading criteria:

**P4** determine the resultant and equilibrant of a system of concurrent coplanar forces from given data;

**M2** describe the conditions required for the static equilibrium of a body.

# Friction

Engineers and designers consider friction all the time. Friction is why cars have rubber tyres and ceramic brakes, why snowboards are made from smooth composites, why running shoes have specially designed soles and why aircraft are designed to be aerodynamic. It is also why your hands get warm if you rub them together and why plant-maintenance and automotive technicians need to replace bearings, gaskets and seals from time to time.

**Figure 4.25** Friction is why snowboards are made from smooth composites

Friction exists whenever two or more materials come into contact with each other. The material does not necessarily need to be solid. Friction from liquids and gases is also a primary concern for engineers and designers, especially those working in the process and energy sectors.

To understand friction properly, it is worth introducing a few scientific terms applied to a simple situation.

If we placed a book on a table, it would have weight. We know from Newton's third law that any action must have a reaction, so the table would produce a reaction force (Figure 4.26). The reaction force acts up on the book (at 90 degrees to the table) – we call this the normal (reaction) force ($F_n$). If the book weighed 10 newtons, then the normal force would also be 10 newtons. If you put another book of the same weight on top, then the normal force would increase to 20 newtons, and so on.

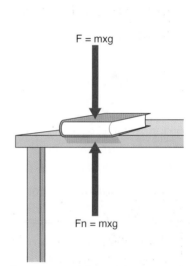

$F = mxg$

$Fn = mxg$

**Figure 4.26** The table produces a reaction force, which acts up on the book

Now, if you tried to push the two books with a finger, you would probably find it relatively difficult to start with, but it would then get easier as the book begins to slide across the table. The reason for this is that the force of friction always opposes the direction of motion; in other words, friction goes the opposite way to the direction the object is travelling – the frictional force attempts to resist the movement.

A small force applied to the book (a tiny shove, for example) will be equalled and opposed by the *static frictional force*. This is the frictional force produced to resist the movement of a stationary object. However, once the book is moving, the frictional force is reduced – we call this the *kinetic (or sliding) frictional force*. The reason for the smaller kinetic frictional force opposing the object when it is in motion is due to the surface roughness and chemistry of the object's material.

## Activity

Attach an elastic band to a mug full of water and pull the mug towards you using the elastic band. Initially, you will notice the elastic band stretching more as the pulling force attempts to overcome the static friction force. Once the mug starts to move, less force will be required to keep it moving and the elastic band will stretch less to overcome the kinetic friction force.

Under close examination, no surfaces are ever perfect (although some materials are better than others) and it is this surface imperfection that causes friction in solid objects. Initially, as the two materials come into contact, they become 'wedged' into the various contours, peaks and valleys of the materials' microscopic surfaces. As such, it takes quite a bit of effort (or force) to get the surfaces to initially move. Once they do move, the surfaces ride over the imperfections and less effort is required to maintain the movement. (Imagine if you were to start pedalling a mountain-bike over a dirt track – if the ground were particularly uneven, it may be a struggle at the start, but once moving you would not feel the uneven ground as much. This is essentially the same principle.)

The effects of friction can be reduced in several ways: machining, casting, moulding or forming materials with better surface properties; using materials that can achieve better surface finish (such as polymers and composites); or using lubricants such as oil and grease. Lubricants work by effectively filling in the peaks and valleys of the surface texture to allow the contact material to move more freely over them, thereby reducing damage through heat and contact wear.

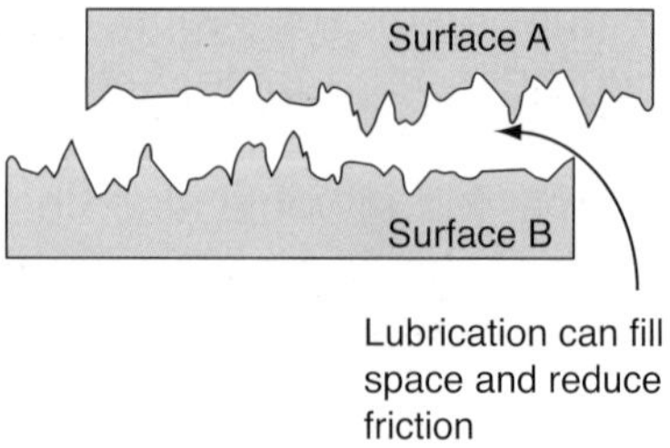

Lubrication can fill space and reduce friction

**Figure 4.27** Magnified surfaces

## The coefficient of friction

The coefficient of friction is a number (normally between 0 and 1) that is used to illustrate how much friction exists between two surfaces. In other words, it is how much 'slip' or 'grip' exists between two materials.

## Activity

Try and find out three different values for the coefficient of friction between two materials. For example:

- metal on metal
- rubber on ice
- wood on snow.

To calculate the coefficient of friction depends on whether the object is moving. As mentioned previously, static friction occurs before an object is moving (for example, this is the force that would prevent a child slipping down a toy slide if wearing a textile with a high coefficient of friction) and kinetic friction occurs when an object is moving (this is what would cause a moving object to eventually slow down and stop moving).

If we know the weight of an object, then we can calculate the normal force ($F_n$).

On a flat surface, weight = normal force = mass $\times$ gravity

The coefficient of friction is calculated using the following equations:

$$\text{coefficient of static friction} = \mu_s = \frac{\text{frictional force (static)}}{\text{normal force}}$$

$$\text{coefficient of kinetic friction} = \mu_k = \frac{\text{frictional force (kinetic)}}{\text{normal force}}$$

## Worked Example

**A block of polystyrene packaging is resting on a steel work table. The coefficient of friction between the two surfaces is 0.35 and the mass is 0.5 kg. What is the static frictional force?**

Solution:

To calculate the answer, we need to rearrange the formula, so:

$$\text{coefficient of static friction} = \mu_s = \frac{\text{frictional force (static)}}{\text{normal force}}$$

Becomes:

Frictional force (static)   = coefficient of static friction $\times$ normal force

= 0.35 $\times$ 4.905 (normal force = mass $\times$ gravity = 0.5 $\times$ 9.81 = 4.905 N)

= 1.72 N

## Make the grade

The information above provides a key foundation in achieving the following grading criterion:

**D2** **determine the work done and the power dissipated in moving a body of given mass along a horizontal surface at a uniform velocity, given the value of the coefficient of kinetic friction between the contact surfaces. We will return to this criterion on page 159**

# Linear motion

Linear means straight and motion means movement – put them together and you get 'straight movement'. The term 'dynamic' is also occasionally applied to linear motion – although some care needs to be exercised when using this term, as it has much wider applications in physical science (for example, angular motion and chemical reactions, which are beyond the scope of the BTEC First in Engineering).

You get linear motion when an F1 car races down the home straight and when an athlete runs down the 100-metre track at the Olympics. You also get linear motion when you walk down the street in a straight line and when an air ambulance flies overhead, direct to the scene of an accident.

To understand linear motion, it is helpful to understand the difference between a number of key terms. You should also understand the difference between a scalar and a vector. If you are still uncertain, refer back to the Team Talk on page 141.

## Distance and displacement

Distance is a measurement between two points of interest. You will no doubt be familiar with many well-used units associated with distance, such as miles, kilometres, centimetres and millimetres, but as far as science is concerned the SI unit is the metre. Distance is a scalar quantity, so it does not have any direction.

Displacement is also a measurement between two points of interest, although being a vector quantity, it is always specified with a direction, for example 30 metres south or 40 metres acting at an angle of 45 degrees.

## Speed

Speed is used to determine the amount of distance a moving object covers in a certain time and is calculated as:

$$\text{speed} = \frac{\text{distance (in metres)}}{\text{time (in seconds)}}$$

The unit for speed is metres per second (m/s).

Although it is acceptable to use kilometres per hour, you need to ensure you are careful with the units. For example, if your car (moving relatively quickly) covered 1.2

## Key words

**Distance** is a measurement between two points of interest.

**Displacement** is also a measurement between two points of interest, but it is always specified with a direction.

**Speed** is used to determine the amount of distance a moving object covers in a certain time.

kilometres in 60 seconds, you should not calculate the speed as:

$$\text{speed} = \frac{\text{distance}}{\text{time}} = \frac{1.2 \text{ kilometres}}{60 \text{ seconds}} = 0.02 \text{ km/s}$$

This is because km/s is not a standard unit. Instead, you should convert the kilometres into metres (by multiplying by 1,000). So the correct calculation would be:

$$\text{speed} = \frac{\text{distance}}{\text{time}} = \frac{1200 \text{ metres}}{60 \text{ seconds}} = 20 \text{ m/s}$$

Speed is calculated using distance (with no direction); therefore, it is a scalar.

## Velocity

Velocity is also a measure of how fast an object is travelling but, unlike speed, it has a direction. It is therefore a vector and uses displacement in its calculations:

$$\text{velocity} = \frac{\text{displacement (in metres with a direction)}}{\text{time (in seconds)}}$$

The units are the same as speed; the only difference between the two is that a direction will be quoted, for example 30 m/s north.

A method of reminding yourself about the difference between speed and velocity is to compare a car speedometer with a satellite-navigation system. Speedometers only tell you how fast you are travelling without any indication of direction. A sat-nav tells you how fast you are travelling and also in what direction – it therefore tells you the velocity.

## Converting speeds and velocities

Table 4.7 provides you with a guide for quickly (and roughly) converting between miles/hour, kilometres/hour and metres/second. For example, if you want to convert m.p.h. into m/s, take your number in m.p.h., look along the row to find the multiplier (for m/s) and multiply the two together.

> **Key word**
>
> **Velocity** measures how fast and in what direction an object is travelling.

**Figure 4.28** Satellite-navigation systems tell you how fast you are travelling and in what direction (velocity)

| m.p.h. | km/hr | m/s |
|--------|-------|------|
| Number | 1.61 ⟶ | 0.45 |
| 0.62 | Number | 0.27 |
| 2.24 | 3.6 | Number |

Table 4.7 Converting between miles/hour, kilometres/hour and metres/second

## Worked Example

**Convert 60 m.p.h. into m/s.**

Solution:

$$60\,\text{m.p.h.} \times 0.45 \text{ (from table)} = 27\,\text{m/s}$$

**Convert 60 m.p.h. into km/hr.**

Solution:

$$60\,\text{m.p.h.} \times 1.61 \text{ (from table)} = 96.6\,\text{km/hr}$$

## Activity

Have a go at the following:

1. **Convert 18 m/s into km/hr.**
2. **Convert 100 km/hr into m.p.h.**
3. **Convert 70 m.p.h. into km/hr.**

**Now have a go at converting the following land-speed records from m.p.h. into km/hr and m/s.**

| Date | Car | Velocity in m.p.h. | Velocity in km/hr | Velocity in m/s |
|------|-----|--------------------|--------------------|------------------|
| 18th Dec 1898 | Jeantaud | 39.24 | | |
| 12th Jan 1904 | Ford | 91.37 | | |
| 8th Nov 1909 | Benz | 109.65 | | |
| 5th Feb 1931 | Bluebird | 246.09 | | |
| 19th Nov 1937 | Thunderbolt | 312.00 | | |
| 23rd Oct 1970 | Blue Flame | 622.41 | | |
| 15th Oct 1997 | Thrust SSC | 763.035 | | |

Table 4.8 Land-speed records

## Acceleration

We all hear the term 'acceleration' when watching TV programmes like *Top Gear* – for example, 'The Bugatti Veyron can accelerate from 0 to 50 km/hr in approximately 1.5 seconds.' But what does this mean scientifically?

Acceleration is the change in velocity in a specific time or the change of velocity per unit time. Acceleration, like velocity, is a vector.

Acceleration is calculated by:

$$\text{acceleration} = \frac{\text{change in velocity}}{\text{time taken}}$$

### Key word

**Acceleration** is the change in velocity in a specific time or the change of velocity per unit time.

### Activity

Where have we been introduced to accleration already in this unit?

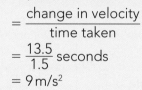
### Worked Example

**If a Bugatti accelerates from 0 to 50 km/hr in 1.5 seconds, calculate the acceleration.**

Solution:

First, we need to sort out the units: the velocity is given in km/hr but the time is given in seconds. The easiest thing to do is convert the 50 km/hr into m/s.

Using Table 4.7 on page 152, 50 km/hr × 0.27 = 13.5 m/s

$$\text{acceleration} = \frac{\text{change in velocity}}{\text{time taken}}$$
$$= \frac{13.5}{1.5} \text{ seconds}$$
$$= 9 \text{ m/s}^2$$

Note: the units for acceleration are m/s²; in other words, metres per second, per second.

To explain the units, it means if the Bugatti had an acceleration of 9 m/s², then it would be travelling 9 m/s in the first second, 18 m/s by the second second and 27 m/s by the third second.

To convert back to where we started (in km/hr), we again use Table 4.7 to convert 27 m/s into km/hr. This means the car is doing 97.2 km/hr (or 60.48 m.p.h.) after only three seconds:

$$27 \text{ m/s} \times 3.6 \text{ (from table)} = 97.2 \text{ km/hr}$$

or

$$27 \text{ m/s} \times 2.24 \text{ (from table)} = 60.48 \text{ m.p.h.}$$

## Activity

1. If a car accelerates from 0 to 15 m/s in 33.7 seconds, calculate the acceleration in m/s².
2. If a car steadily accelerates from 5 m/s to 12 m/s in 30 seconds, calculate the acceleration in m/s².
3. If a car accelerates from 14.5 km/hr to 52.5 km/hr in 24 seconds, calculate the acceleration in m/s².
4. If a car brakes from 12 m/s to 2 m/s in 3 seconds, calculate the deceleration in m/s².

## Force, acceleration and mass

Once we have calculated acceleration, it is very easy to determine the force of an object moving in a straight line. We have come across the following equation already:

force (N) = mass (kg) × acceleration (m/s²).

## Worked Example

**If a motorbike with a mass of 150 kg accelerates at 5 m/s², what is the force?**

Solution:

$$\text{force (N)} = 150 \times 5 = 750\,\text{N}$$

## Distance–time graphs

Distance–time graphs can be used to graphically show the speed of an object in motion. An example is shown in Figure 4.29.

In this case, a few things are of interest:

- Between points A and B, the car covered 10 metres in 2 seconds.
- Between points B and C, the car did not cover any additional distance in 5 seconds.
- Between points C and D, the car covered 5 metres in 3 seconds.
- Between points D and E, the car did not cover any distance in 20 seconds.

You can draw only limited conclusions about what is

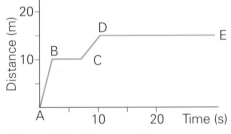

Distance-Time Graph

**Figure 4.29** Distance–time graph

happening to the car from the distance–time graph, but it is possible to determine the speed.

In reality, distance–time graphs have little use because they provide the user with limited detail; more useful are velocity–time graphs, which are used in exciting engineering developments, including rocket engineering, land-speed records and Formula 1 racing.

## Velocity–time graphs

The velocity–time graph may look very similar to the distance–time graph, but it presents different information. Firstly, the slope of the velocity–time graph does not represent the speed (as in the distance–time graph) but the acceleration of the object. Secondly, by calculating the area under the slope, it is possible to determine the distance the object has travelled.

On the velocity–time graph in Figure 4.30, a number of points are of interest:

- Between points A and B, the car increased its velocity from 5 m/s to 8 m/s in 6 seconds. The acceleration is therefore:
  $$\text{acceleration} = \frac{\text{change in velocity}}{\text{time taken}} = \frac{3\,\text{ms}}{6} = 0.5\ \text{m/s}^2$$

- Between points B and C, the car stayed at 8 m/s – it has not stopped; it just has not accelerated.

- Between points C and D, the car accelerated from 8 m/s to 12 m/s in 9 seconds.

- Between points D and E, the car decelerated from 12 m/s to 0 in 10 seconds.

**Activity**

Calculate the following speeds:

1 Between points A and B

2 Between points B and C

3 Between points C and D

4 Between points D and E

The answers are given at the back of the book.

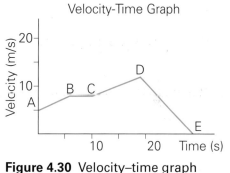

**Figure 4.30** Velocity–time graph

**Activity**

Calculate the acceleration/deceleration between the following points:

1 C and D
2 D and E

The answers are given at the back of the book.

We can also calculate the distance travelled by totalling up the area under the graph.

## Activity

The area under the line represents the distance travelled. Calculate the distance travelled in metres.

The answer is given at the back of the book.

## Make the grade

There are a number of useful websites designed to allow simple conversion between units. Google 'unit conversion' to find a number of examples. The next activity will help you in achieving the following grading criterion:

**P5** determine the uniform acceleration/retardation of a body from given data.

## Activity

The following activity uses a popular games console and a popular racing game. Any racing game can be used as long as it has an option to display the speedometer of the vehicle as it races the circuit (which can be a track or street race).

To complete the activity, follow the steps below:

1. Prepare a table like this:

**Figure 4.31** Racing game

| Time | Velocity in m.p.h. or km/hr | Velocity in m/s |
|------|------------------------------|-----------------|
| 0    |                              |                 |
| 10   |                              |                 |
| 20   |                              |                 |
| 30   |                              |                 |
| 40   |                              |                 |
| 50   |                              |                 |
| 60   |                              |                 |

2. Select a game option that allows you to complete one circuit of the track or street course.

3. Race the car around the circuit and ask a fellow student to collect the time from the game speedometer every ten seconds. It will be given in either m.p.h. or km/hr.

4. Convert the velocity into m/s using Table 4.7 on page 152.

5. Take these times and construct a velocity–time graph. The time should be plotted on the horizontal axis (x axis) and the velocity should be plotted on the vertical axis (y axis).

6. Repeat the exercise with another student driving, and collect their data.

7. Analyse the results. A couple of points of interest should include:

   a) Who achieved the highest velocity?

   b) Who was the quicker driver?

   c) Who was the more economical driver?

This type of activity simulates (to some extent) the type of information F1 engineers analyse when trackside at testing, qualifying and racing sessions. They can use similar information to determine fuel economy, whether the car is performing at the optimum level and whether the driver is performing well.

# Work done and power

D2

## Mechanical work

On a level surface, work = force (newtons) × distance (metres)

$$W = f \times d$$

Therefore, the unit is the newton.metre (Nm), which is re-termed the joule (J) – this is also the unit of energy. To do work, you need energy; energy is a form of stored work. If you pushed two children on a sledge, you would use up energy and produce work.

Two children on a sledge are pushed with a force of 100 newtons. If they were pushed 10 metres, the work would be:

$$W = f \times d$$

work = 100 newtons × 10 metres = 1000 joules or 1 kJ

If you are calculating the work to lift an object, then we replace the force with weight and distance with height. Therefore, the equation becomes:

work to lift an object = weight (newtons) × height (metres)

### Key word

**Work** is the force required to move an object over a distance.

## Worked Example

**If your mass is 80 kg and you jump 10 cm into the air, how much work is done?**

Solution:

First, you need to convert mass into weight:

$$\text{weight} = \text{mass} \times \text{gravity} = 80 \times 9.81 = 784.8 \text{ newtons}$$

Then, you need to convert 10 cm into metres: 0.1 metres.

Now apply the equation:

$$\text{work done} = \text{weight} \times \text{height} = 784.8 \times 0.1 = 78.48 \text{ joules}$$

## Power

Power (P) is calculated using the work (done) and the time taken.

$$\text{power} = \frac{\text{work}}{\text{time taken}}$$

$$P = \frac{W}{t}$$

Power is measured in watts (W) – 1 watt can be defined as how many joules of work (or energy) are used in 1 second.

### Key word

**Power** is a measure of how much energy has been used in a specific period of time.

## Worked Example

**Back to the earlier example, if it took 20 seconds to push a sledge 10 metres with 100 newtons of force, then how much power was used?**

Solution:

$$\text{work} = 100 \text{ newtons} \times 10 \text{ metres} = 1000 \text{ joules}$$

$$\text{power} = \frac{\text{work}}{\text{time taken}} = \frac{1000 \text{ joules}}{20 \text{ seconds}} = 50 \text{ watts}$$

## Activity

1. We previously introduced Newton's laws of motion and throughout the unit we have shown how they consistently underpin the science associated with mechanical engineering. Here is a quick opportunity to revise where we have discussed them. Write down Newton's laws of motion and explain what they mean.

2. Power is calculated as: force $\times$ velocity (for a moving object)

Can you recall another method of calculating power?

## Make the grade

The next activity will help you in achieving the following grading criterion:

**D2** determine the work done and the power dissipated in moving a body of given mass along a horizontal surface at a uniform velocity, given the value of the coefficient of kinetic friction between the contact surfaces.

## Activity

3. A 3 kg block of steel resting on a cast-iron marking table is pushed along by an engineer at a velocity of 0.75 m/s for 1.3 seconds. If the coefficient of kinetic friction for steel and cast iron is 0.23, calculate:

- the work done;
- the power dissipated in moving the sledge.

# Learning Outcome 1. Be able to define and apply concepts and principles relating to electrical science

## The 'First' guide to electrical science

The second strand to the science element of the First Diploma in Engineering is the introduction of the electrical principles.

Electrical science is what underpins everything from our domestic electricity supply, providing our homes with light and heat and powering the numerous gadgets and appliances we take for granted: media players, games consoles, computers, TVs, DVD players and Blu-ray players, as well as washing machines, dishwashers, toasters and fridges.

Like gravity in mechanical science, electrical theory is largely based on a quantity that can barely be seen but nevertheless can be harnessed to carry out a multitude of tasks to help us at home, at work or with our hobbies and leisure activities. Electrical science has a colourful history of scientists and engineers who have been highly influential in the development of understanding in the field. One such person is Georg Ohm, whom we will introduce briefly in this part of the unit through 'Ohm's law', which remains arguably the most fundamental law of all electrical theory.

## The atom

If you had to remember just one important piece of information about science, then arguably that should be that all things are made of atoms.

It takes trillions of atoms to make up you and everything around you. Atoms even make up things you cannot see, such as the air you breathe.

A simple search on the internet will tell you that a 70 kg person will have approximately $7 \times 10^{27}$ atoms of matter such as oxygen, hydrogen and carbon.

Two or more atoms make up a molecule, and the ability of atoms and molecules to join groups is essentially the building block of everything in the universe. The Sun, stars and planets, like us, are all made of atoms.

The standard illustration of an atom (the one you are likely to have been introduced to at school) is termed the Bohr model (named after Danish physicist Niels Henrik David Bohr). It consists of a nucleus (containing protons and neutrons) surrounded by orbiting electrons (Figure 4.32).

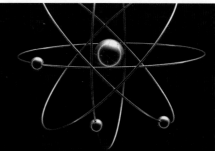

**Figure 4.32** The Bohr model

## Conductors, insulators and ions

In some materials (conductors), the atoms contain electrons that can be easily dislodged from their orbits around the nucleus. This may happen as a result of heat, light, vibration or friction. These electrons are called 'free electrons' and any atom that loses an electron would become positively charged (as it would have more of the positively charged protons remaining) – this is termed a positive ion. Any material that gains electrons (and therefore has more negatively charged electrons than protons) is termed a negative ion.

The party trick of rubbing a balloon on your hair and then sticking it to the ceiling is a result of this transfer of electrons (which charges the balloon).

- Materials with a high number of free electrons are termed conductors.
- Materials with a low number of free electrons are termed insulators.

## Current and electromotive force

The direction of the free electrons in a material is random and, unless they are effectively pushed in the right direction, they will not have much use. The force used to push the electrons is called electromotive force, or e.m.f.; more commonly, it is referred to as voltage. Electromotive force is measured in volts (V).

The flow of electrons is called electric current and is measured in amperes, commonly abbreviated to the amp (A).

## Charge

As ions are charged either positively or negatively, it is important to have a unit to provide an indication of how much charge they posses. The unit of charge is the coulomb (C). A single electron has a tiny charge of $1.6 \times 10^{-19}$ C.

Charge and current are closely related by time:

charge = current $\times$ time

Q (coulombs) = I (amps) $\times$ t (seconds)

### Activity

**How can the formula be rearranged to make current the subject?**

### Worked Example

**A digital camera is left on charge via a USB cable for one hour. If the current is 0.7 A, what is the charge?**

Solution:

$$Q = I \times t$$

$$Q = 0.7 \times 3600$$

(Note the time in seconds is 1 hour $\times$ 60 minutes $\times$ 60 seconds = 3600 s)

$$Q = 2520\,C$$

**Figure 4.33**

## Power

Like mechanical power, the unit for electrical power is the watt and both are a measure of the amount of energy that is transferred in a given time. Specifically, electrical power can be derived using the following formula:

power = voltage $\times$ current

P (watts) = V (volts) $\times$ I (amps)

## Worked Example

**If a Dyson vacuum cleaner ran on a domestic supply of 230 V and produced 1,250 W of power, what would be the current?**

Solution:

$$\text{power} = \text{voltage} \times \text{current}$$

So, rearranging the formula gives:

$$\text{current} = \frac{\text{power}}{\text{voltage}}$$

$$= \frac{1250}{230}$$

$$= 5.43 \text{ A}$$

**Figure 4.34**

## Resistance

It was Georg Ohm, a German physicist and schoolteacher, who discovered in 1827 that there was a direct relationship between voltage and current.

He observed that an increase in voltage resulted in a proportional increase in current. He also noted that it varied from material to material.

He concluded that some materials restrict the flow of electrons more than others and termed this phenomenon 'electrical resistance (R)', giving his name to the unit, the ohm (with the symbol Ω). An easy way to understand electrical resistance is to compare it to a garden hose – if the water flow through the hosepipe is compared to the flow of electrons (current), then by squeezing the hosepipe, you will restrict (or resist) the water flowing through (resistance).

Understanding and applying resistance is essential in the design of any electrical or electronic system. If you took any modern appliance apart and examined the printed circuit board (PCB), you would see a scattering of components like those shown in Figure 4.35 – resistors. Without these components, the current and associated voltage entering into highly sensitive components, such as integrated circuits and microprocessors, would cause excessive load and catastrophic damage. The resistors act as gates, designed to allow just the right amount of current through.

What is now known as Ohm's law is used to calculate voltage, resistance and current when only two of the quantities are known.

$$\text{Voltage} = \text{current} \times \text{resistance}$$
$$V(\text{volts}) = I(\text{amps}) \times R(\text{ohms})$$

**Figure 4.35** Resistors

## Activity

**Rearrange Ohm's Law to make**

a   **the current; and**

b   **the resistance the subject.**

## Worked Example

An electrical circuit has 110 volts of voltage and 3 amps of current. What is the resistance?

Solution:

$$R = \frac{V}{I}$$

$$= \frac{110}{3}$$

$$= 36.67 \ \Omega$$

## Make the grade

The next activity will help you in achieving the following grading criterion:

**P1** define parameters of **direct current electricity and magnetic fields.**

## Activity

**Table 4.9 contains some mistakes – try and identify them.**

| Term | Symbol | Formula | Standard Unit |
|---|---|---|---|
| Charge | Q | $Q = I \times t$ | coulomb (C) |
| Electromotive force | E | $V = I \times R$ | volt (V) |
| Current | I | $I = V \times R$ | amp (A) |
| Resistance | R | $R = V / I$ | watt (W) |
| Power | P | $P = V / I$ | ohm ($\Omega$) |

**Table 4.9**

# Direct current circuits (series)

P2

The simplest electrical circuits allow you to switch on a light at home; some of the most complex allow you to download music and games from the internet.

In basic terms, an electrical circuit will have:

- a source of e.m.f. (normally from a battery or domestic mains supply);
- a load (components such as resistors or lamps);
- a switch;
- current flow.

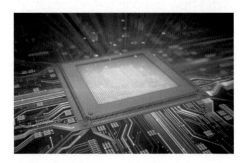

**Figure 4.36** Printed circuit board

Conventional current flow is generally used on circuit diagrams showing the current flowing from the positive terminal of the supply to the negative. In reality, actual current (or electron) flow goes the other way (as the

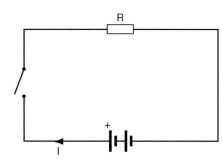

Figure 4.37 Basic electrical circuit

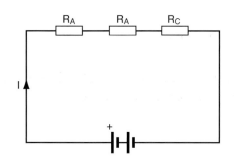

Figure 4.38 Resistors in series

negatively charged electrons move towards the positive terminal).

When the resistors in a circuit are connected end to end, they are said to be in series (Figure 4.38).

As the current passes through each of the resistors, depending on the resistance, the component will experience a 'voltage drop' or 'potential difference'. Effectively, this means that the force used to push the electrons around the circuit gets exhausted by the resistance of the components.

## 66 Team Talk

Aisha: **'How do you measure current?'**
Steve: **'It's quite easy, really. You can use a device called a multimeter. The multimeter can be used to measure voltage, current or resistance. Most electricians carry one as an essential part of their toolkit.'** 99

In a circuit such as the one in Figure 4.39, the resistors have an equal value. If the e.m.f. supply were 110 volts, then the potential difference in each resistor would be 36.67 volts ($\frac{110 \text{ volts}}{3 \text{ resistors}}$).

If the values were not the same, then Ohm's law could be used to calculate the potential difference, as long as the current was provided. It is very important to note that in a series circuit the current is the same, no matter where it is measured.

For example in the circuit shown in Figure 4.40, the current is 0.44A.

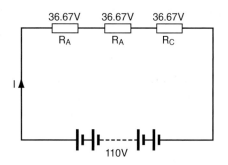

Figure 4.39 Circuit

In the 100 ohm resistor, the potential difference is:

$V = I \times R$

$V = 0.44\,A \times 100\,\Omega$

$\quad = 44$ volts

In the 80 ohm resistor, the potential difference is:

$V = I \times R$

$V = 0.44\,A \times 80\,\Omega$

$\quad = 35.2$ volts

In the 70 ohm resistor, the potential difference is:

$V = I \times R$

$V = 0.44\,A \times 70\,\Omega$

$\quad = 30.8$ volts

As a check, 44 volts + 35.2 volts + 30.8 volts = 110 volts (which was the original supply voltage)

You can calculate the total resistance in a series circuit by either:

● using Ohm's law if the current and total voltage are known; or

● adding the values of the individual resistors together.

Total resistance (in ohms) = resistance of A + resistance of B + resistance of C

$R_T = R_A + R_B + R_C$

The total resistance can be calculated by:

$R = \dfrac{V}{I}$

$\quad = \dfrac{230}{5}$

$\quad = 46\,\Omega$

or:

$R_T = R_A + R_B + R_C$

$\quad = 12 + 18\ 1\ 16$

$\quad = 46\,\Omega$

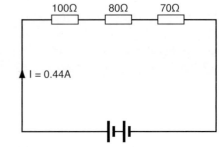

**Figure 4.40**

# Direct current circuits (parallel)

The second type of circuit is the parallel circuit, which has a number of differences from the series circuit introduced in the preceding section.

In the parallel circuit, the current effectively splits between the resistors, causing different values of current in each component depending on the resistance. The potential difference in a parallel circuit is the same across each resistor connected in parallel regardless of the resistance.

Using Ohm's law, it is possible to calculate the current in each resistor, if we know the supply voltage and resistor values.

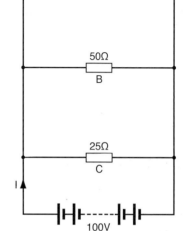

**Figure 4.41** Parallel circuit

So, in Figure 4.41:
the current in resistor A:

$$I = \frac{V}{R}$$
$$= \frac{100}{25}$$
$$= 4 \text{ A}$$

the current in resistor B:

$$I = \frac{V}{R}$$
$$= \frac{100}{50}$$
$$= 2 \text{ A}$$

the current in resistor C:

$$I = \frac{V}{R}$$
$$= \frac{100}{25}$$
$$= 4 \text{A}$$

The total current is therefore 4 + 2 + 4 = 10 A.

The total current can now be used to calculate the total resistance by applying Ohm's law once again:

$$R = \frac{V}{I}$$
$$= \frac{100}{10}$$
$$= 10\,\Omega$$

An alternative method of calculating the total resistance (useful if the current or voltage is unknown) is as follows:

$$\frac{1}{R_T} = \frac{1}{R_A} + \frac{1}{R_B} + \frac{1}{R_C}$$

Using the previous example:

$$\frac{1}{R_T} = \frac{1}{25} + \frac{1}{50} + \frac{1}{25}$$

$$\frac{1}{R_T} = 0.04 + 0.02 + 0.04$$

$$\frac{1}{R_T} = 0.10$$

Therefore, rearranging gives us:

$$R_T = \frac{1}{0.10} = 10\,\Omega$$

## Make the grade

The next activity will help you in achieving the following grading criterion:

**P2** determine total resistance, **potential difference and current in series and parallel DC circuits from given data.**

## Activity

1. Three resistors are connected in series. Resistor A has a value of 120 ohms, resistor B has a value of 19 ohms and resistor C has a value of 82 ohms. If the supply voltage is 230 volts, calculate:

   a) the total resistance;

   b) the current;

   c) the voltage drop across each resistor.

2. The same three resistors are connected in parallel. Calculate:

   a) the current across each resistor;

   b) the total current;

   c) the total resistance.

3. Four resistors are connected in series. Resistor A has 5 ohms, resistor B has 7.2 ohms and resistor D has 12.9 ohms. If the total resistance is 23 ohms and the current is 3 amps, calculate:

   a) the resistance in resistor C;

   b) the supply voltage;

   c) the voltage drop across each resistor.

**P1** **M1**

# Magnetism

Like gravity and voltage, magnetism is an invisible, yet very powerful, force that can be used by scientists and engineers in products we use every day. The ignition of a car engine, automatic door locks and a variety of sensors and motors all work because of magnetism and electro-magnetism.

You will no doubt have come across a magnet before. Perhaps a bar made from iron. Iron is used as it has excellent magnetic properties and is commonly used in many of the applications mentioned previously.

The bar magnet always has a north and south pole which can be used to attract or repel other similar bar magnets. As you will probably know, two different poles attract and two similar poles repel each other. A further common lesson in school is to use iron filings around the magnet to show the magnetic field and magnetic flux.

## The magnetic field, flux and flux density

The magnetic field is the area around the magnet where another ferromagnetic object will be influenced. This means any ferromagnetic object will either be attracted or repelled by the magnetic force.

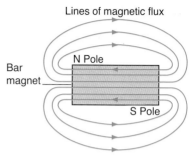

**Figure 4.42** Magnetic lines of flux

 Team Talk

Aisha: **'What is ferromagnetic?'**
Steve: **'This means a material made of iron, or a material that acts like iron in terms of its magnetic properties. Nickel and cobalt are two such materials. Some metals such as aluminium and copper are not magnetic at all.'**

The magnetic flux (phi or $\Phi$) is a term used to describe the lines of magnetism and therefore the strength of a magnetic field surrounding the magnet. The lines will always be shown diagrammatically as running from north to south.

- If the lines of flux are very close together, you have a high magnetic flux and therefore a strong magnetic field.

- If the lines are further apart, you have a lower magnetic flux and therefore a weak magnetic field.

To describe the strength of a magnetic field, the magnetic flux has its own unit – the weber (Wb).

## Magnetic-flux density

Some care needs to be exercised when discussing 'magnetic flux' and 'magnetic-flux density' as the two terms have slightly different meanings.

While magnetic flux ($\Phi$) refers to the spacing of the lines of flux, magnetic-flux density (B) refers to the lines of flux in a specific area. Magnetic-flux density also has a different unit: the tesla (T).

As the two are related by area, the formula is:

$$\text{magnetic-flux density} = \frac{\text{flux}}{\text{area}}$$

$$\beta \text{ (in tesla)} = \frac{\Phi \text{ (in weber)}}{A \text{ (in metres}^2)}$$

To generate one tesla requires a magnetic field to generate one newton of force from only one amp of current.

## Conductors and force in magnetic field

If an electrically conducting material has a current passed through it, it will generate a magnetic field, as shown in Figure 4.43. If more current is passed through the conducting material, the magnetic field will become stronger.

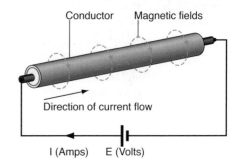

**Figure 4.43** Current carrying conductor

In Figure 4.43, you can see the lines of flux around the conductor flowing in a clockwise motion. If the current were travelling the other way, then the lines of flux would move anti-clockwise. We'll come back to this in a later section.

To calculate the force generated on the conductor in a magnetic field, the following formula can be used:

force on a conductor = magnetic    × current × length of
                        flux density                conductor

F (newtons ) = B (tesla) × I (amps) × I (metres)

This relationship is extremely important – it is the basis for many motors, solenoids and sensors.

## Activity

1. Sketch out the bar magnet shown in Figure 4.44 and draw in the lines of flux.
2. What are the units for magnetic flux and magnetic-flux density?

**Figure 4.44** Bar magnet

## Activity

If a conductor in a magnetic field experiences a force of 3.74 newtons and the length of the conductor is 46 millimetres, calculate the magnetic-flux density required if 3 amps are passed through the circuit.

# The construction and use of electro-magnetic coils

**D1**

An electro-magnetic coil is formed when a current-carrying conductor (such as a copper wire) is wound round a material known as a 'core' to form a coil. When a current is passed through the conductor, the coil generates a magnetic field (Figure 4.45).

## The right-hand grip rule

It is possible to determine the direction of the field (i.e. north or south) and the direction of the current using the right-hand grip rule. The thumb points to the north pole (which is generated as a result of the direction of current) and the fingers point in the direction of the current (Figure 4.46).

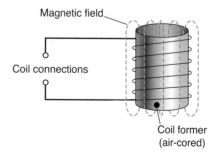

Magnetic field

Coil connections

Coil former (air-cored)

**Figure 4.45** Electro-magnetic coil

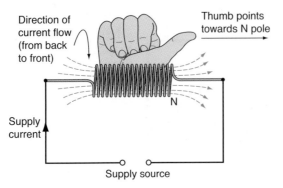

**Figure 4.46** The right-hand grip rule

## Strength of the electro-magnetic coil

It is possible to increase the magnetic-field strength of a coil by:

- increasing the current passing through the conductor;
- increasing the number of turns around the core;
- varying the type of core material – an iron core will provide a much stronger magnetic field than a copper or polymer version.

## Uses of electro-magnetic coils

Electro-magnetic coils have a variety of uses. Some examples are described below.

### Solenoid

Solenoids are very simple electro-magnetic coils that are commonly used in switches and valves. The coil surrounds a conducting material and when the coil is energised (by applying an electric current), the conductor moves into or out of the coil, producing a linear force. This force can be used to actuate a switch or valve.

One of the most common uses of a solenoid is in the ignition of a car engine. When you turn the key, a small current is sent to the solenoid, which is used to make a closed-circuit contact between the car battery and the starter motor, which turns the engine over.

Other common uses for solenoids are on modern door locks that are opened using photo ID or key fobs. Once the fob or photo ID is swiped through a sensor, a small current is sent to the solenoid, which energises the coil and moves the core to unlock the door.

Solenoids have many uses in industry, where they are used to control production lines. Even some paintball guns work by using solenoids.

## Relays

There are a number of different types of relays, as they have many uses in electrical and electronic circuits. A relay is an electrical switch that uses electro-magnetism to move a hinged ferromagnetic link (known as an armature) to close or open a circuit contact. The electro-magnetism comes from a coil surrounding a soft iron core. When a current passes through the coil, the resulting magnetic attraction moves the armature, thereby switching the circuit. When the current is removed, the armature returns to its original position.

## Sensors

Sensors are used in systems to detect the presence of physical quantities. Sensors using electro-magnetism are often used in control technology as inputs into programmable logic controllers (PLCs) used to control highly efficient production and assembly lines. Sensors are also used in computer technology for hard drives and in cars as the basis for speedometers and anti-lock braking systems (ABS).

## Transformers

Transformers are used to increase or decrease the voltage entering into an electrical or electronic device. For example, your mobile phone requires approximately 5 volts to charge the lithium battery. The transformer takes the 230 volts as supplied from the mains and 'transforms' it into the 5 volts that the phone requires. It does this by using two electro-magnetic coils surrounding a core.

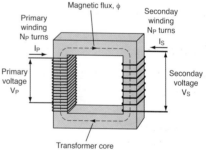

**Figure 4.47** Transformer

- If the number of turns were the same on both sides, then the transformer would have no influence on the circuit (i.e. the voltage 'in' would equal the voltage 'out').
- If the turns on the primary side were lower than the secondary side, then the transformer would step-up (increase) the voltage leaving the device.
- If the turns on the primary side were higher than those on the secondary side, then the transformer would step-down (decrease) the voltage leaving the device, as is the case with the car battery charger.

# Activity

**Use the internet and other sources to research the answers to the following questions.**

1. **Investigate the use and construction of a direct current (DC) motor.**
2. **What is a commutator?**
3. **What is the purpose of the brushes?**
4. **What is an armature?**

## Make the grade

The detail in this section will help you in achieving the following grading criterion:

 **D1** **explain the construction, function and use of an electro-magnetic coil.**

---

## ℹ️ Grading criteria recap

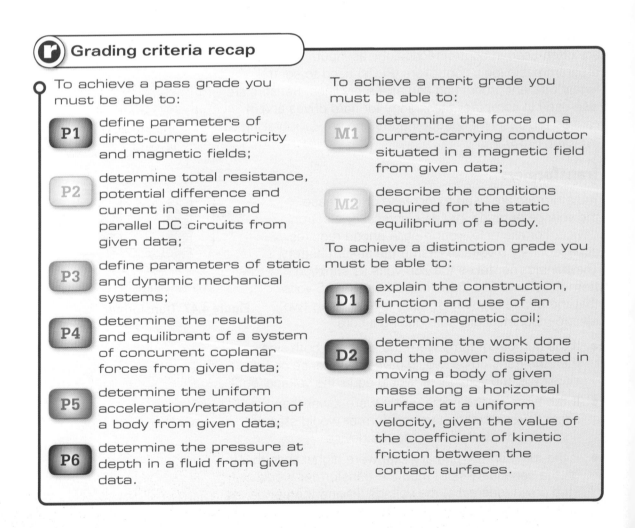

To achieve a pass grade you must be able to:

**P1** define parameters of direct-current electricity and magnetic fields;

**P2** determine total resistance, potential difference and current in series and parallel DC circuits from given data;

**P3** define parameters of static and dynamic mechanical systems;

**P4** determine the resultant and equilibrant of a system of concurrent coplanar forces from given data;

**P5** determine the uniform acceleration/retardation of a body from given data;

**P6** determine the pressure at depth in a fluid from given data.

To achieve a merit grade you must be able to:

**M1** determine the force on a current-carrying conductor situated in a magnetic field from given data;

**M2** describe the conditions required for the static equilibrium of a body.

To achieve a distinction grade you must be able to:

**D1** explain the construction, function and use of an electro-magnetic coil;

**D2** determine the work done and the power dissipated in moving a body of given mass along a horizontal surface at a uniform velocity, given the value of the coefficient of kinetic friction between the contact surfaces.

# Introduction to the unit

Engineering materials are selected for a wide range of products, from typical engineering structures such as bridges and skyscrapers, to modern hand-held gadgets such as MP3 players and mobile phones. This unit will give the learner the opportunity to understand why different materials are used in engineering by looking at two important topics: the different properties of engineering materials and how different materials are identified.

Everyone has some knowledge of common engineering materials, but this unit will take that understanding further by explaining how materials work and why they are chosen, using examples and applications from different areas of engineering. A basic understanding of materials will provide an excellent platform for other engineering units at this level or when progressing to the next level of study.

## Learning Outcomes

By the end of this unit you should:

- understand the properties of common engineering materials;
- know how engineering materials are identified.

# Grading criteria

| To achieve a pass grade you must be able to: | To achieve a merit grade you must be able to: | To achieve a distinction grade you must be able to: |
| --- | --- | --- |
| **P1** describe the properties that are used to define the behaviour of common engineering materials | **M1** explain the choice of material for a given engineering component based on the material's properties | **D1** establish that a material has the required properties for a given application |
| **P2** review the properties of a given ferrous metal and a given non-ferrous metal | **M2** select an appropriate form of supply for a given material requirement | |
| **P3** review the properties of a given organic material, a given thermoplastic, a given thermosetting polymer and a given smart material | | |
| **P4** identify symbols and abbreviations used on given engineering documentation | | |
| **P5** identify the forms of supply available for a given engineering material | | |

# Learning Outcome 1. Understand the properties of common engineering materials

## Putting materials into categories

Engineering materials can be put into different categories, which is an advantage to an engineer as it makes life easier when selecting a material. At the start of this unit, the categories are very basic, but you will notice that they split into sub-categories as the unit progresses.

The categories that will be used at this early stage are:

- metal;
- polymer;
- wood;
- ceramic;
- composite (a material that is constructed from two or more different materials).

You may be surprised at how well you do at the following activity – already you will have some knowledge of common engineering materials used in everyday life.

## Activity

Fill in the columns in Table 8.1. Under each heading, write down examples of each type of material (some examples have been provided). You could take turns in a group to name an engineering material and write it into the correct category. A group of 12 or more students should be aiming for the following total number of materials:

| | |
|---|---|
| 35–40 | Average |
| 40–45 | Good |
| 45–50 | Very good |
| 50+ | Excellent |

This exercise will determine your prior knowledge of engineering materials at the start of this unit. To help you, an example of a material is included for each category. (Hint: if you are struggling, look around the room or out of the window – most materials will fit into one of these categories.)

| Metal | Polymer | Wood | Ceramic | Composite |
|-------|---------|------|---------|-----------|
| Copper | PVC | Pine | Porcelain | Fibreglass |
| | | | | |
| | | | | |
| | | | | |
| | | | | |
| | | | | |
| | | | | |
| | | | | |

**Table 8.1** Putting materials into categories

# Physical and mechanical properties of materials

P1

When selecting a material, it is important to consider the properties of that material. To the novice, it could be misleading to describe a material as strong, hard or tough – you could be forgiven for thinking that they mean the same thing. However, these properties are completely different. We will now describe these (with examples), along with a range of other basic properties.

## Activity

**What properties do you think are important for a hand-held games console? Complete Table 8.2 by thinking of four more properties from any part of the console. (Hint: think of what is inside the console as well as what you can see on the outside.)**

**Figure 8.1** Hand-held games console

| Property | Reason |
|----------|--------|
| Toughness | So that it does not crack or break when dropped |
| | |
| | |
| | |

**Table 8.2** Important properties for a hand-held games console

# Physical properties

## Density

Density can be described as the mass per unit volume of a material. A frisbee is usually made from plastic, but imagine what would happen if you made a frisbee with exactly the same dimensions but from steel. The sizes would be the same but the densities would be different because steel is heavier and more dense than plastic. Here is another way to understand density: imagine you are in a swimming pool, timing how long it takes to walk one width. Now imagine the pool is empty and you walk the same width. It would be a lot quicker when the pool is empty because water is denser than air.

**Figure 8.2** A steel frisbee would be denser than a plastic frisbee

## Melting point

The melting point is the temperature when a material changes from a solid into a liquid. The melting point is crucial for engineers when they select a material for a certain application, especially where heat is involved (e.g. ovens, barbeques, etc.).

### Activity

Choose ten different materials from Table 8.1 and find out their melting points. This can be done as an individual exercise (competing against each other) or as a team activity against the clock. Use either books or the internet to complete the activity.

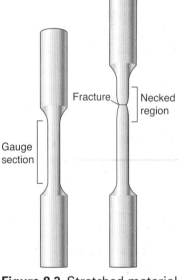

**Figure 8.3** Stretched material after a tensile test

# Mechanical properties

## Strength

Strength can be split into sub-categories. Two of the main sub-categories are tensile strength and compressive strength.

*Tensile strength:* if you apply enough force to a material, then you can change its shape. Heavy loads can stretch a material and are used in testing to ensure that certain materials are suitable for selection. Tensile strength is the ability to withstand heavy loads. Figure 8.3 shows how a metal can extend (in this case, from a tensile test). In this

### Key word

**Tensile strength** is the ability to withstand heavy loads.

example, the metal becomes longer but also thinner in the middle; this is referred to as necking.

A good example of tensile strength can be seen on a steel wire rope on a crane. The wire rope is designed to lift a safe working load (SWL). After prolonged use, weathering and possible misuse by exceeding the SWL, the wire rope stretches by a small degree. The rope is inspected on a regular basis and is changed when it stretches too far. This is because the diameter of the rope decreases and it is more likely to snap with a heavy load. If you stretch a piece of Play Doh, you will see that, as it extends, the middle becomes thinner and you need less load to snap it.

**Figure 8.4** When a car crashes, the body dents and changes shape

*Compressive strength:* this is the ability of a material to withstand compressive loads or resist being squashed. Concrete has an excellent compressive strength so is used on structures such as bridges (because these structures need to resist loads from thousands of vehicles). If a material such as sandstone were used, it would break and crumble under the load (as it possesses a poor compressive strength).

## Toughness

Toughness is the ability of a material to withstand a sudden shock or load. When a car crashes and is subjected to impact loads, the body of the car (usually made from steel) dents and changes shape because it is made from a tough material.

## Brittleness

The opposite of toughness is brittleness. A brittle material cracks, snaps or fractures instead of changing shape. Glass is a very brittle material and can only take impact loads when strengthened. Human bones are also brittle, which results in a lot of sports-related injuries.

## Hardness

Hardness is the ability of a material to withstand scratching, abrasion or indentation. If you use a scriber to draw a line across a piece of wood (see Unit 18

Engineering Marking Out), the material not only shows a line but also becomes indented. If you drew a line across a piece of steel, it would only scratch the surface. This is because steel is a much harder material than wood. Hardness can be measured in various ways and using basic experiments; this will be discussed at a later stage.

## Activity

Diamond is the hardest natural material known to man. It is commonly used in jewellery but is also used extensively in engineering. Find three engineering uses of diamond.

## Ductility

Ductility is the ability of a material to change shape without breaking. The shape of some materials is changed so they can be used for a particular purpose. A car body that is made from low-carbon steel can be pressed into shape and does not break because it is a ductile material. Copper rods can be pulled through dies (small holes) to reduce the diameter; after the process is repeated several times, the rods end up as copper wire or tubes. This process is known as drawing and is possible because copper is also a ductile material.

**Figure 8.5** Copper is a ductile material

## Malleability

Malleability is similar to ductility – it describes the ability of a material to change shape without cracking or rupturing. However, a malleable material plastically deforms after being subjected to a compressive force (such as hammering or squeezing). Rivets are made from malleable materials and are used as a permanent fixture to keep two or more pieces of material together – this technique is commonly used on aeroplanes. The small circles in Figure 8.5 show how the sheet metal is permanently held together.

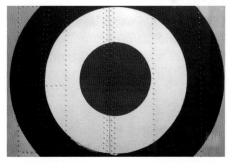

**Figure 8.6** Rivets on an aircraft

## Elasticity

Elasticity is the ability of a material to return to its original shape after being stretched or compressed. The obvious example of this is an elastic band as

**Figure 8.7** Some materials display excellent elastic properties

it displays excellent elastic properties. An elastic band can be stretched to several times its own length and still return to its original shape. It may be surprising to hear that some metals have good elastic properties. Once a material goes past its 'elastic limit', it will not return to its original shape. If you bend a clear plastic shatterproof ruler far enough, it will develop a white line and will no longer lie flat. This is because you have bent it past its elastic limit. The degree of elasticity of a material can be worked out in a tensile test (see Figure 8.3). Tensile strength and elasticity are properties that are closely linked.

## Why mechanical properties are important

If you use the example of a crane mentioned earlier, then you can see how some of the mechanical properties already discussed are essential for use in industry. We are already aware that a wire rope has to have good tensile strength so that it does not stretch and snap under load. Figure 8.8 shows how a wire is situated around a crane.

The vertical part of the wire rope with the hook on the end is subject to tensile loads. However, the wire rope needs to travel around a pulley wheel on the top of the crane and then wind around a drum at the back of the cab. As the rope is coiled around the drum, more properties need to be considered.

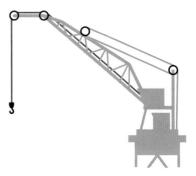

**Figure 8.8** Wire rope on a crane

The wire rope must be *ductile* enough to be able to wind around the drum. It must not be too *brittle* as it would fracture and break on the pulley wheel and the drum. It also has to be *tough* enough to take the impact that comes from loading cargo. Wire ropes also have a degree of *elasticity* so that any sudden loads will not snap the rope.

## Activity

**Give an example where a product would need three or more mechanical properties.**

**This can be used in assignment work or as a learner presentation.**

# Electro-magnetic, chemical and durability properties

P1

## Electro-magnetic properties

### Electrical conductivity

Electrical conductivity is when a material is able to allow an electric current to flow through it. Copper is commonly used in wire form to conduct electricity; aluminium is also widely used. Gold is also an excellent conductor of electricity and is used in electrical contacts as it does not corrode easily. However, the best conductor of all metals is silver! Precious metals such as gold and silver are not commonly used as wires due to their expense.

## Electrical resistivity

Electrical resistivity is the opposite of conductivity. A material with this property opposes the flow of the electric current. Materials possessing good electrical resistivity are referred to as insulators. Glass is one of the best insulators, but most types of plastics are commonly used to insulate wiring due to their low cost and ease of manufacture.

## Ferromagnetism

Magnets are very common; it is likely that you will have some on your fridge door holding notes. Materials that retain their magnetism are known as permanent magnets. These are usually made from iron, nickel and cobalt. Soft iron can be used as a material for soft magnets. Soft magnets are only magnetic when placed in a magnetic field; they lose their magnetism when the field is removed. An example of this is a pile of nails. They are not magnetic but touch one with a magnet and it will pick the other nails up.

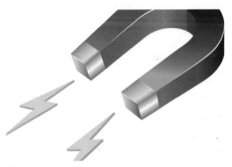

**Figure 8.9** Permanent magnets are usually made from iron, nickel and cobalt

## Chemical and durability properties

### Resistance to corrosion

If a material has a low resistance to chemical attack, then it can easily corrode. Some materials have a natural resistance to corrosion, such as copper, aluminium and most plastics. Copper does not react with water but does react with oxygen; this results in the build-up of a layer of copper oxide, which prevents corrosion.

Aluminium also builds up an oxidised layer to prevent corrosion, but it is not as visible as in copper. This is because the aluminium-oxide layer is very thin and is the same natural colour as aluminium.

## Environmental degradation

A common type of corrosion that you may be familiar with is rusting. Rusting occurs when iron is subjected to moisture (or water) and reacts with oxygen. Corrosion can be reduced or prevented by coating with a material such as zinc or by applying liquid polymer (painting).

**Figure 8.10** The Statue of Liberty

Plastics can also degrade as they can discolour or become more brittle. This can be attributed to factors such as UV or visible light, temperature, oxygen or impurities from manufacture. You may have witnessed this with the yellowing of door and window frames or on casings for computers and kitchen appliances.

## Durability

If a material can resist 'wear and tear', then it is said to be durable. Some companies use this term as a marketing tool to promote their products (batteries and paint are promoted to be long-lasting and hard-wearing respectively). Durable materials are usually needed when two or more surfaces continuously come into contact.

# Thermal properties

## Thermal conductivity

If heat flows through a material freely, then the material is said to be a good thermal conductor. Copper and stainless steel are commonly used to make cooking pans as the heat from the hob conducts through the base to heat the contents. If the pan's handle is also made from stainless steel and water is left to boil in the pan, then heat travels (conducts) along the handle, which can sometimes be too hot to hold. To avoid this, a wooden or plastic handle can be used because wood and plastic are not good thermal conductors. The disadvantage of this is that you would not be able to put the pan under the grill or in the oven as the wood or plastic may burn.

## Thermal expansion

Varying temperatures can affect a material's size (length, height or thickness). This is known as thermal expansion. Engineers need to take this into account when they design structures such as bridges or railways to ensure that slight changes in size due to thermal expansion do not damage the structure. If you listen to the noise in a car when you drive over a bridge, or listen to the noise on a train, you will notice the irregular sound when passing over expansion joints. These are used to stop materials from buckling or kinking by leaving a gap ready for expansion in the warmer months. The term used for this is 'sun kink'.

# Optical properties

## Transparent

If you can clearly see through a material, it is said to be transparent. The reason for this is because it transmits light. Glass is an excellent example.

## Opaque

If you cannot see through a material, it is said to be opaque. Opacity occurs when the material does not transmit light. Wood and steel are good examples.

## Translucent

A material that lets only some light through is said to be translucent. This occurs by the scattering of light through the material. Frosted glass and tracing paper are good examples.

## Make the grade

This activity will help you in achieving the following grading criterion:

**P1** describe the properties that are used to define the behaviour of common engineering materials.

## Activity

Give a description of the properties shown in Table 8.3, which are used to define the behaviour of common engineering materials.

| Material property | Description |
|---|---|
| **(Tensile) strength** | E.g. the ability of a material to withstand tensile (stretching) loads without breaking. |
| **Hardness** | |
| **Elasticity** | |
| **Toughness** | |
| **Ductility** | |
| **Electrical conductivity** | |
| **Electrical resistivity** | |
| **Ferromagnetism** | |
| **Environmental degradation** | |
| **Thermal conductivity** | |
| **Thermal expansion** | |

**Table 8.3** Properties used to define the behaviour of common engineering materials

# Common engineering materials

We have discussed some of the different categories of materials at the beginning of the unit. Figure 8.11 shows how different groups of materials can be split up into categories.

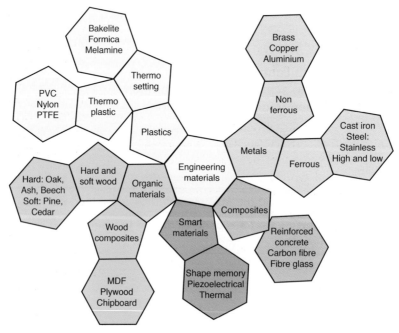

**Figure 8.11** Classification of materials

## Activity

Look around a room in your house and see what things are made of. Why are they made from certain materials and what alternatives could be used?

# Ferrous materials

## Ferrous metals

Metals that contain iron are said to be ferrous. The ferrous metal that you will be most familiar with is probably steel. Steel also contains non-metal carbon and it is the amount of carbon present in steel that dictates its inherent properties. The actual content of carbon in steel is very low. Carbon steel is usually split into three categories: low, medium- and high-carbon steel.

## Low-carbon steel

Low-carbon steel (containing approximately 0.05–0.29 per cent carbon) is also referred to as mild steel. It is the most commonly used plain-carbon steel because it has a relatively low cost. Its other main characteristics are that it is a relatively tough, malleable, ductile material with a good tensile strength that can be easily formed. However, it does have a low resistance to chemical attack, which means that it corrodes easily. It is used to make chains, pipe, wire and rivets, and for general engineering purposes. It has a shiny, silver appearance.

## Medium-carbon steel

Medium-carbon steel (containing approximately 0.3–0.79 per cent carbon) has a high tensile strength and is also tough and hard. It manages to balance ductility with strength, so it can be used where both properties are desirable. We discussed a good example earlier in this unit: a wire rope on a crane. Other uses include gears, crankshafts, hammers and screwdrivers. Medium-carbon steel also reacts to heat treatment, changing its properties for specific purposes. It has a dark silver appearance.

## Key words

**Ferrous** is commonly used to describe metals that contain iron.

**Alloys** contain two or more elements that have been mixed together to change or enhance properties. Alloys contain a metallic element mixed together with either another metallic/non-metal element or a chemical compound and are commonly used in engineering today.

**Carbon steel** refers to steels that change their properties as carbon content is added in small proportions.

**Synthetic** is something that is made artificially.

**Composition** refers to the different elements that make an alloy (similar to ingredients in a cake).

## High-carbon steel

High-carbon steel (containing approximately 0.8–1.4 per cent carbon) is a very hard steel but is also brittle. Toughness is quite low but this can be increased with heat treatment. High-carbon steel has very good wear resistance. It is used on many hand tools, such as chisels, files and saws, so hardness and wear resistance are key properties. High carbon steel has a darker grey colour than steels with lower carbon content.

**Figure 8.12** Steel spring

## Stainless steel

Stainless steel is another common ferrous metal. This is an alloy of iron, carbon, chromium, nickel and magnesium. It was mentioned earlier that alloys are made to change properties and that the main disadvantage of steel is that it corrodes easily. By adding the extra elements to the steel, it becomes non-corrosive (does not rust) and can be used to make cutlery, pans and surgical instruments. Stainless steel has a bright silver, shiny appearance.

## Cast iron

Cast irons are also ferrous metals and contain more carbon than the plain-carbon steels. Grey cast iron has approximately 3.4 per cent carbon and can be used on bulkier products, such as engine blocks, fly-wheels and machine beds. It used to be a popular metal for bridge building. Cast iron tends to be wear resistant, with good rigidity, but can be brittle. Cast irons tend to have a silver-grey appearance.

## Activity

Carbon steel containing 0.3 per cent carbon (low-carbon steel) can be used to make car bodies, whereas carbon steel containing 0.6 per cent carbon (medium-carbon steel) can be used to make chisels. Explain why this is the case, highlighting in your answer the materials' properties.

# Non-ferrous materials

P2

## Non-ferrous metals

Metals that do not contain iron are called non-ferrous metals; they can be pure metals or alloys. You will probably have heard of the materials in this section, such as copper and aluminium, as they are commonly used in engineering.

### Copper

Copper is commonly used in engineering as it is an excellent electrical and thermal conductor. In fact, the only other metal that is better than copper in these areas is silver. It is also a ductile, malleable metal with relatively high strength; that is why it can be drawn into wire form quite easily. Another advantage of copper is that it can be machined and permanently joined by soldering, brazing or welding.

Copper can be used for a wide range of products, including wires, pipes, printed circuit boards, cookware and as a component of coins. Copper changes appearance depending on its state. In its pure form, it has a bright metallic, almost pink, colour, but it can also have a green appearance after reacting with oxygen. Copper is 100 per cent recyclable and it is estimated that 80 per cent of all mined copper is still in use today.

### Aluminium

Bauxite is the main ore from which aluminium is obtained. Aluminium is the most widely used non-ferrous metal and most of the products that are referred to as aluminium are actually aluminium alloys. This is because pure aluminium has a low tensile strength, but mechanical properties can be improved with alloying. Aluminium and aluminium alloys are good thermal and electrical conductors, have low density, are lightweight and have excellent resistance to corrosion. Aluminium is used on aeroplanes, cars, bicycles, kitchen equipment and electrical wires; in fact, the list goes on. Aluminium has a silver metal appearance, which can be slightly dull due to surface oxidation.

## Activity

Using Table 8.4, list what you think is the most important property for each of the five applications of aluminium given in the previous paragraph.

| Application | Aeroplane | Car | Bicycle | Kitchen equipment | Electrical wires |
|---|---|---|---|---|---|
| **Property** | E.g. lightweight | | | | |

**Table 8.4** Aluminium applications

## Brass

Brass is an alloy of copper and zinc. The properties of brass largely depend upon the amount of zinc in the alloy. However, brass is a malleable material and one of its common uses that you are probably aware of is in musical instruments. One of the main reasons brass is used for musical instruments such as trumpets and trombones is because of its acoustic properties. Brass is also quite hard and a good conductor of heat and electricity. Another common use is on door decorations, such as door handles and street numbers, mainly because of its ornamental appearance and anti-corrosive properties. Brass has a yellow and sometimes gold-like appearance depending on its composition.

## Bronze

Bronze is an alloy of copper and tin. You will have seen bronze medals being awarded for finishing third in a race but it is not a precious metal such as gold or silver. As well as copper and tin, bronze can also be alloyed with phosphorus, aluminium, manganese or silicon, depending on the required properties. In the past, it has been used to make weapons and for decorative purposes (see later in this unit on the history of materials). Its properties depend on the composition of the bronze but it is generally stronger and has better resistance to corrosion than brass. Bronze has a red-yellow appearance but the colour can change slightly with composition.

## Lead

Lead is a pure metal that is extremely malleable, very soft and heavy. Over the years, it has been used in many applications because of its useful properties, but in more recent times it has been used less because of environmental reasons. It is still used in car batteries (lead-acid battery) and in the building trade for roofing material and gutters. One of the disadvantages of lead is that it is poisonous and, with long-term exposure, can affect major organs or the nervous system. Environmental waste is another reason why lead is starting to be replaced with more modern materials. Lead has a dull bluish-grey appearance.

> **Make the grade**
>
> This activity will help you in achieving the following grading criterion:
>
> **P2** review the properties of a given ferrous metal and a given non-ferrous metal.

## Activity

We have now discussed the properties of a range of ferrous and non-ferrous materials. Using the tables below, give a rating out of 10 (10 being excellent and 1 being very poor) for the properties of mild steel (Table 8.5) and aluminium (Table 8.6) and give reasons for your ratings (you could include examples of typical applications).

| Ferrous metal – mild steel | | |
|---|---|---|
| Property | Rating out of 10 | Reason for rating |
| Strength | | |
| Hardness | | |
| Toughness | 9 | A tough material that does not break easily; this is one of the reasons it is used for car bodies |
| Ductility | | |
| Resistance to corrosion | 1 | Will rust, especially if used outside. Needs to be coated – for example, paint on mild-steel handrails |
| Electrical conductivity | | |
| Ferromagnetism | | |
| Thermal conductivity | | |
| Durability | | |

**Table 8.5** Ferrous metal rating

**Non-ferrous metal – aluminium**

| Property | Rating out of 10 | Reason for rating |
|---|---|---|
| Strength | | |
| Hardness | | |
| Toughness | | |
| Ductility | 9 | A ductile material that can be processed in various ways. Used on aircraft parts as it is easily machined and formed |
| Resistance to corrosion | | |
| Electrical conductivity | 9 | A very good conductor of electricity and is commonly used in power transmission lines |
| Ferromagnetism | | |
| Thermal conductivity | | |
| Durability | | |

**Table 8.6** Non-ferrous metal rating

# Polymers

P3

Polymers are commonly known as plastics. They are synthetic materials that are often used because of their low cost. They can be split into three main categories: thermoplastics, thermosetting polymers and elastomers.

## Activity

Arrange the following polymers into three columns and include one application of each. The headings for the three columns are:

- **Thermoplastics**
- **Thermosetting plastics**
- **Elastomers**

| | | |
|---|---|---|
| Nylon | Epoxy resin | PVC |
| Rubber | Acrylic | Polyester |
| Bakelite | ABS | Melamine |
| Polythene | Polybutadine | PTFE |

## Thermoplastics

Thermoplastics can be heated up, melted and re-formed into another shape. This is ideal for recycling unwanted items such as plastic bottles and packaging. Figure 8.13 shows how a thermoplastic can be reshaped once it has been melted. Alternatively, if a thermoplastic is frozen, it can be changed into a glass-like state. Before they are set, thermoplastics are in a granular, powder or pellet form and are mixed with a colouring agent and filler to determine the appearance and properties of the plastic.

**Figure 8.13** Thermoplastics can be reshaped

Some examples of thermoplastics are as follows.

### ABS (acrylonitrile butadiene styrene)

ABS is very useful as it is a hard, tough, strong, durable plastic that can be used on casings for computers, mobile phones and MP3 players, as well as for engineering applications on automotive parts and casings for electrical goods. It is a versatile plastic because of its many properties and also because of its resistance to corrosion and good surface finish (smooth).

### PVC (polyvinyl chloride)

PVC can be used in two different forms. You may have heard of PVC clothing, which is made from a soft, flexible material. The same type of plastic is also used as a sheath or covering on electrical cables. It is flexible because plasticizers have been added to change the properties.

PVC that is un-plasticized is referred to as uPVC. The properties of uPVC are different due to the fact that the plastic becomes hard, tough and stiff. It is commonly used on window frames, drainpipes and gutters because of its properties and low cost.

PVC is recyclable, but there are concerns about the harmful emissions of PVC, which could impact on the future use of the plastic.

### Activity

Use the internet and other sources of information to compile a short paragraph on the harmful emissions created by PVC and uPVC.

### Nylon (polyamide)

Nylon is commonly used in engineering as well as in fabrics. It famously replaced silk in the manufacture of parachutes in the Second World War because silk was in short supply. In its harder form, it is used for engineering applications, such as gears, bearings and hinges because its properties include high durability, self-lubrication and good resistance to chemical attack. Nylon is a hard, tough polymer and can change in appearance from a white to a creamy colour, or it can even have a translucent appearance (see 'Optical properties' on page 185).

### PTFE (polytetrafluoroethylene)

PTFE is more commonly known by the trade name Teflon®. You probably have heard of this as a coating on non-stick frying pans and other cookware. PTFE is also used in electrical applications because it is an excellent insulator and in the food and pharmaceutical industries because it can resist high temperatures. It is also used in bearings, bushes and gears because it has low-friction properties. The appearance of PTFE is usually white to light grey with a very smooth surface finish.

### Polythene (or polyethylene)

Polythene is the most commonly used plastic and should be familiar to you as the plastic shopping bag. Its properties vary depending on whether it is low-density polythene (LDPE) or high-density polythene (HDPE). LDPE is a flexible, tough material and quite soft; its applications include plastic bags, packaging (e.g. the plastic rings that hold four cans of lager together) and computer components, such as disc drives. HDPE is harder, stiffer and has a higher tensile strength than LDPE. Its applications include tables, chairs, water pipes and milk bottles.

## Thermosetting polymers

Thermosetting polymers are moulded into shape and cannot be reshaped. This is because a chemical change takes place as they are initially heated. This change is called polymerisation and, once it takes place, the plastic stays in a hard and rigid state. An easy way to remember the difference between thermoplastics and thermosetting polymers is by recognising that 'set' appears in the word to describe the types of plastics that can be moulded only once: thermo**SET**ting polymers.

A good analogy for this is to think of boiling an egg. An egg is in a liquid state before cooking. When heat is applied during boiling, the egg becomes set into a solid state. Once the egg is set into a solid state, it cannot be melted or re-formed, in the same way as a thermosetting polymer. If you continue to heat the egg (or the thermosetting polymer), it will burn or decompose.

Some examples of thermosetting polymers are as follows.

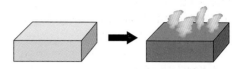

**Figure 8.14** Thermosetting polymers can burn or decompose

## Epoxy resin

Epoxy resin is formed by two different chemicals; hence the name (a material formed by two different chemicals is also called a copolymer). You may have experienced something like this if you have ever mixed a paste and hardener together to form a plaster, whereby the chemicals emit a slight amount of heat (chemical reaction).

**Figure 8.15** Once set, an egg, like a thermosetting polymer, cannot be melted or re-formed

## Bakelite®

Bakelite® was the first synthetic plastic and was developed by Dr Leo Baekeland in 1909. It is a very poor conductor of electricity and heat so can be used as an insulator. It was mainly used for applications such as telephone casings, plug sockets and other electrical goods. As technology has moved on and 'greener plastics' have become available, Bakelite® products have become collectable items. There is even a museum in Somerset dedicated to Bakelite®.

## Formica®

Formica® is a heat-resistant, plastic laminate of paper or fabric mixed with melamine resin, which makes it a type of composite. It is manufactured with layers of paper, which are held together with the resin before a protective layer is added to the surface for protection. It is then compressed (squashed) and heated to harden the surface. Formica® is a brand name from the American Formica Corporation.

## Melamine

Melamine is a hard plastic that is made by polymerisation. It is used as the main constituent in Formica®, but it is also used in its own right as a plastic. It does tend to scratch easily so is usually quite cheap. Its applications include kitchenware and camping utensils. It is also used in panels for kitchen cabinets and self-build furniture goods to give a hard, clean surface.

## Polyester resin

Polyester resin is commonly used in a liquid form to produce composites such as glass-reinforced plastics. It produces a hard, stiff plastic that has good resistance to chemical attack and can be used as an electrical insulator. It is used in the manufacture of car bodies, castings and as an adhesive.

## Reinforcing materials

Reinforcing materials such as glass fibres, carbon fibres and wood flour all contribute to the manufacture of composite materials. They are used along with materials such as polyester resin to create a material with specific properties for a certain application.

*Glass-reinforced plastic* (commonly named as fibreglass) can be used for applications such as tent poles, fishing rods and pole-vault poles because it is lightweight, flexible and strong.

*Carbon fibres* are combined with a plastic resin to form a composite material. Carbon fibre-reinforced polymers are commonly used in engineering because of their excellent properties. They are used on applications such as aircraft, boats, motorcycles and cars because they are strong and lightweight. They are also a cheaper alternative to aluminium or titanium, which have similar properties.

*Wood flour* is wood that has been finely ground into small particles so that it can be used as filler in thermosetting polymers such as Bakelite®. Wood flour is likely to be a constituent in the decking used in the gardening industry. In fact, it has probably been used in most plastic-wood building products. High-quality wood flour is made from hardwood, whereas the cheaper wood flour is made from softwood so is of a poorer quality.

> **" Team Talk**
>
> Aisha: **'How are composites made?'**
> Steve: **'There are different methods of making composites and a lot of them involve some sort of binder (e.g. adhesive or resin) mixed with a reinforcement.'**
> Aisha: **'Give me two examples, then.'**
> Steve: **'Fibreglass and carbon-fibre composites are made with long fibres of glass or carbon (reinforcement) that are set in resin (binder) to form materials with good properties.'** "

## Elastomers

Elastomers are polymers with the property of elasticity (discussed earlier in this unit, using the example of a rubber band). Elasticity is the most common characteristic of elastomers. They are used on applications such as tyres and seals because considerable, reversible extensions are possible.

# Organic materials and smart materials

P3

## Organic materials

There are three main categories of organic materials used in engineering: hardwoods, softwoods and wood composites.

### Hardwoods

After learning about the properties of materials, you may think that a hardwood has the property of hardness. However, this is not true for all hardwoods. The reason they are in this category is because they are harvested from angiosperm trees (mostly deciduous, which means that they are broad-leaved and lose their leaves in the winter). In fact, some hardwoods are softer than some softwoods – balsa wood is a good example. Balsa wood is very soft; you could make an indentation by crushing it with your fingers. Despite this, it is classed as a hardwood. Hardwoods are categorised by the seeds they produce. Hardwoods produce seeds with a covering, such as a fruit (apple, pear) or a hard shell (acorn).

There are many more types of hardwoods than softwoods, and these include:

- oak;
- ash;
- beech;
- maple;
- teak;
- mahogany;
- ebony.

## Softwoods

Softwoods come from coniferous trees, referred to as evergreens because they do not lose their needles in the winter (conifers have needles rather than leaves). Softwoods are used in building materials much more than hardwoods because they are cheaper and tend to be easier to work with. They grow faster than hardwoods, which is why they are cheaper. As well as for building materials, softwoods are used for furniture and in the manufacture of paper.

Types of softwoods include:

- pine;
- cedar;
- redwood;
- spruce.

## Wood composites

Wood composites are sometimes referred to as man-made woods and are manufactured to enhance the properties of natural wood. Wood fibres, particles or veneers (thin layers of wood) are bound together with adhesives so that a manufacturer can produce a material that meets a particular specification. There are several types of wood composites available. Two examples are as follows:

- Plywood – veneers are bonded together at right angles to each other so that the wood is stronger. It is very easy to snap a thin piece of wood, but if you have two pieces or more on top of each other, with the grains running in opposite directions, then it would be difficult to snap! Adhesives, usually in the form of resins, are used to bind the wood.
- Medium-density fibreboard (MDF) – MDF is a cheap

material that is made by mixing wood flour (or sawdust) with resin and pressing it into the shape of a board.

## Smart materials

Smart materials are materials that can alter their properties according to changes in their surroundings, such as temperature, stress or by passing an electrical charge through them. This means that they can respond to situations in a positive manner without human interference. It is likely that you will have already seen or used a smart material. Some of these are detailed below.

**Figure 8.16** Smart materials are used in LCDs on mobile phones

## Chromogenic systems

Chromogenic systems can be split into three categories, as follows:

- Thermochromic materials – these change their colour at a particular temperature. You may have seen this in cups or children's bath mats.
- Photochromic materials – these change their colour depending upon the brightness of the light. These materials are used in light-sensitive sunglasses.
- Electrochromic materials – these change their colour when a voltage is applied. You will have seen this in liquid crystal displays (LCDs) on mobile phones, computers and televisions.

## Piezoelectric materials

Piezoelectric materials are smart because they can produce a small electrical discharge when a material is deformed (the word 'piezo' is Greek and means 'to squeeze'). They can also change in size if an electrical current is passed through them (by expanding or contracting). This makes them ideal for use as sensors, for example, in car airbags. The airbag is deployed when the impact from a crash is sensed by the material, which in turn sends an electrical signal to deploy the bag. Another interesting use is on the pick-up of an electro-acoustic guitar where the sound waves bend the material and cause a change in voltage. Quartz is the most common piezoelectric material.

### Activity

Quartz is one of the most common crystals on Earth. Use the internet and other sources of information to find five uses for quartz.

## Shape-memory alloys (SMA)

Shape-memory alloys are smart because they can return to their original shape once heat is applied. They are lightweight and can be used in place of more traditional methods of actuation, such as hydraulic and pneumatic systems. The three main types of alloys used for SMAs are copper–zinc–aluminium–nickel, copper–aluminium–nickel, and nickel–titanium.

An ideal application for SMAs is on aircraft where SMA wires can be used to control a bent wing surface by heating the wires with an electrical current, as opposed to using extensive hydraulic systems to operate the flaps on the wings. This makes the aircraft lighter, more compact and easier to maintain.

### Activity

A good 'seeable' example of SMAs is wire spectacle frames which return to their original shape after being deformed by applying heat. Type 'memory metal' into an internet search engine to find videos showing interesting applications of shape memory metals.

## Magneto-rheostatic and electro-rheostatic fluids

These are smart fluids because they change their viscosity (thickness) from a thick liquid to an almost solid state once exposed to a magnetic or electric field. The advantage of this method is that the reaction is reversible once the field is removed. Minute iron particles that are suspended in oil can be used as a magneto-rheostatic fluid. Electro-rheostatic fluids include cornstarch, oil and even milk chocolate!

### Make the grade

This activity will help you in achieving the following grading criterion:

**P3** review the properties of a given organic material, a given thermoplastic, a given thermosetting polymer and a given smart material.

### Activity

List three key properties of the materials in Tables 8.7–8.10 and describe why they are important. Include applications where possible to back up your answers. An example answer is given in Table 8.9 to help you.

| Organic material – pine | |
|---|---|
| Key property | Description of key property |
| 1. | |
| 2. | |
| 3. | |

**Table 8.7** Organic material

| Thermoplastic – polyvinyl chloride (PVC) | |
|---|---|
| Key property | Description of key property |
| 1. | |
| 2. | |
| 3. | |

**Table 8.8** Thermoplastic

| Thermosetting polymer – Bakelite® | |
|---|---|
| Key property | Description of key property |
| 1. Electrical insulation | Does not conduct electricity so is used as an insulator in applications such as plugs and sockets |
| 2. | |
| 3. | |

**Table 8.9** Thermosetting polymer

| Smart material – quartz | |
|---|---|
| Key property | Description of key property |
| 1. | |
| 2. | |
| 3. | |

**Table 8.10** Smart material

# Testing materials

D1

The testing of materials is a very important part of engineering. A good example of this is in bridge building where the bridge must be strong enough to take enough weight (or force) so that it does not collapse. Years ago, large engineering structures such as bridges and skyscrapers were 'over-engineered' so that they would last the test of time. More recently, engineers have used a factor of safety to help design structures to eliminate over-engineering (computers have played a major part in this). Destructive tests are carried out on materials to prove that they would be strong enough to be used on the structures.

## Activity

Search for 'West Point Bridge Designer 2007' on the Internet, download it and install. It is a free competition-based software pack that enables you to design bridges and test them in 3D.

Other destructive tests are used on different applications to test for toughness, hardness or fatigue. Even your training shoes will have gone through tests to check for wear and tear!

We are going to look at informal testing at this stage and concentrate on investigating the hardness of materials. Informal testing provides an insight into the properties that some materials have over others.

Before this test is carried out, the following health and safety guidelines should be followed:

- Safety goggles should be worn throughout the test.
- Correct PPE should be worn in the workshop (boots and overalls).
- Only use the hammer and punch on a soild work surface.

## Activity

**Informal hardness test**

You can use any materials you want, but try and use materials with the same dimensions (5 mm thick plate, 50 mm × 50 mm, would be useful). Use four materials for the test (e.g. mild steel, aluminium, Perspex® and copper). You can estimate what you think the result will be before you start, rating the materials from hardest to softest.

Carry out three basic tests on each material and score each material for each test.
Test 1: Use a hammer and centre punch to create an indentation on all four materials (try to ensure that the same amount of pressure is applied each time). Use a magnifying glass to identify the material with the largest indentation. The material with the largest indentation is the softest (1 mark) and the material with the smallest indentation is the hardest (4 marks). Award 2 and 3 marks to the materials in the middle.

Test 2: Use a coarse emery cloth and rub the surface of the material, again applying the same amount of load. The material with the roughest scratches is the softest (1 mark) and the material least affected is the hardest (4 marks).

Test 3: Place each piece of material in a vice and use a hacksaw to cut through it. Get a friend to time you and see how many seconds it takes. It is important that you apply the same amount of weight on the hacksaw each time for consistency. The material that is the quickest to saw through is the softest (1 mark) and the material that takes the longest time to saw through is the hardest (4 marks).

Collate your results in a table, as shown in Table 8.11. The material with the highest total is the hardest.

This test compares your selected materials for hardness only.

| Material | Test 1 | Test 2 | Test 3 | Total |
|---|---|---|---|---|
| Mild steel | | | | |
| Aluminium | | | | |
| Perspex® | | | | |
| Copper | | | | |

**Table 8.11** Results of the hardness test

## 66 Team Talk

Aisha: 'How do you carry out formal hardness tests so that results can be used in industry?'
Steve: 'One way of determining a material's hardness is to use a diamond or steel ball indenter on a hardness testing machine. The machine uses weights to ensure that the same force is applied with each test. It also has a microscope to measure the diagonal distance across the flats left by the diamond pyramid or the diameter of the steel ball's indentation. These tests are called the Vickers test (diamond) and Brinell test (steel ball).' 99

Figure 8.17 shows the effects of a Brinell hardness test on a material.

## Activity

**Informal impact test**

Once you have finished with the samples from the hardness test, you can use them for an informal impact test.

The correct PPE should be used in this test and the test should only be carried out under the supervision of an experienced practitioner.

Tighten approximately a quarter of each specimen in a vice so that the rest of the material is protruding from the top. Hit the material with a hammer and make a note of what happens. An impact test measures toughness (see 'Properties of materials' on page 180). You can see how tough the material is by how it bends, dents or snaps. Make a note of your findings. (The materials that bend 90 degrees show that they are ductile and tough. The ones that crack or break lack toughness.)

As an alternative test, put a piece of pipe over the sample and apply a bending force.

## Formal impact testing

This is carried out with material specimens that are notched (have a 'v' cut into them) and then are clamped into testing apparatus for an Izod test (Figure 8.18). The material is subjected to an impact load from a swinging pendulum (the same swing is always used to offer consistency). When the pendulum strikes (impacts) the material, a measurement is taken on a dial to see how much energy has been absorbed. The unit used for this test is the kilonewton (kN). The material will either bend or break depending on its toughness. A lot can be learned from the material after the test, such as how it fractures or tears. You can also look at the grain structure under the microscope.

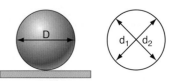

**Figure 8.17** Brinell hardness test

Non-destructive testing (NDT) is used on objects that cannot be damaged. Once an aircraft is built, it goes through a series of non-destructive tests to find out if it is safe to fly without damaging it in the process. NDT is commonly used in engineering, particularly in welding. Some of the processes include the following:

- Dye penetrant – also known as liquid penetrant testing, a dye is used to highlight any cracks in a material.
- Ultrasonic – sends pulse waves through a material to check thickness or for deformations.
- Radiographic – penetrates materials in a similar way to an X-ray.

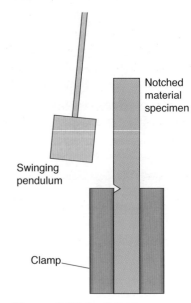

**Figure 8.18** Izod test

- Magnetic particle – used to test steel for defects by adding a magnetic field and using small metal particles. The particles are attracted to the defects.

# History of materials

Engineering materials have been used for thousands of years, but in ancient times only a small range of materials was available. However, with the spread of industry, demands of warfare and the advent of space travel, advances in materials technology have left us with a wide range to choose from.

## Engineering materials timeline

### Before 3000 BC – the Stone Age

Ceramics and natural materials such as stone, clay, wood and flint were used for earthenware. Gold and copper were used for ornamental use and jewellery, while meteoric iron was used for tools and weapons.

### 3000–2000 BC – the Bronze Age

Bronze (an alloy of copper and tin) was an important material; it was used for weapons, armour and tools, even as a building material.

**Figure 8.19**

### 1800–1 BC – the Iron Age

Iron smelting took place (melting iron to produce a metal from iron ore) to produce steel. In this period, glass was used as a sharp tool before the Egyptians used it for jewellery. It was later used for windows by the Romans as the glass industry became more popular. Concrete was used by the Egyptians and the Romans.

*Note – the Stone, Bronze and Iron Ages differ in years depending on regions. Approximate years have been used in this case.*

### AD 1–1000

Porcelain was invented, better known as china, and used for crockery. Some of the properties that made porcelain attractive were its strength, toughness and translucency.

## 1000–1500

Type metal alloys (used in the printing industry), which were made from tin, lead and antimony, were developed. These alloys revolutionised printing in this era.

## 1500–1800

Glass lenses were developed for use on microscopes and telescopes. Techniques for manufacturing steel called 'crucible steel' were developed.

## 1800–1900

The thermocouple (a device similar to a thermometer, which measures voltage related to temperature difference) was invented by Thomas Seebeck. Metallic aluminium was first produced in 1825. Vulcanised rubber was patented by Charles Goodyear. This changed the properties of rubber, making it more durable, harder and resistant to chemical attack.

## 1900–1940

In the First World War, technological advances were practically compulsory and the use of materials to develop applications such as telephones, armoured vehicles, aircraft and wireless communications were much needed. Bakelite® was invented in 1909 and stainless steel was invented in 1912 (used in aircraft engines in the First World War). Glass-reinforced plastic was invented in 1938 and was used mainly as a home-insulation product.

## 1940–1960

Nylon was first produced in 1935, but it was during the Second World War that it became widely used. It replaced silk on parachutes and was also used to manufacture tyres and other military supplies. Shortly after the war, superalloys were developed to provide higher mechanical strength and resistance to high temperatures. This led to advances in the manufacture of military electric motors, gas turbines (aircraft and marine) and submarines.

## 1960–1980

The space age began when the USA and Soviet Union competed with each other in what was termed the 'space race'. This accelerated technology, and new materials were tried and tested for satellites, rockets and modern advances in computers. In this age, materials were designed to suit the required function. The constituents of materials and the way they were formed allowed scientists to create materials with specific microstructures to meet the desired properties.

### 1980–2000+

Optical fibres were tested over a long period, dating back to the 1950s; however, they were not widely used until the early 1990s. Optical fibres are made from glass or plastic fibres and can be as thin as a human hair. They carry digital information (light to produce images or sound across telephones) over long distances at high bandwidths and have the advantage of resisting electro-magnetic interference, unlike metal wires. Rare earth materials, such as iron neodymium boride, were developed. These materials are used as magnets that are of a very high quality. They are used on applications such as sensors, electrical motors on cars and computer disc drives.

## Activity

**Using the information provided, as well as the internet, produce a timeline poster, presentation or model of furnaces throughout history.**

# Learning Outcome 2. Know how engineering materials are identified

## Symbols and abbreviations

You may be aware already that abbreviations and symbols are used for engineering materials. The actual chemical symbols can be found in the periodic table. The symbols that are likely to appear in this unit are as follows:

| | | |
|---|---|---|
| Cu – copper | Fe – iron | Ni – nickel |
| Au – gold | Pb – lead | Al – aluminium |
| Sn – tin | Ti – titanium | Zn – zinc |
| Ag – silver | W – tungsten | Cr – chromium |

Sometimes symbols can be combined to identify certain types of alloys, but these do not appear in the periodic table. For example, 'Sn/Pb' which is tin/lead (also known as solder). Other materials are abbreviated on engineering drawings and do not exist in the periodic table, such as STL, an abbreviation for steel.

There are many symbols and abbreviations found on engineering drawings and these are discussed in detail in Unit 2 (Interpreting and using engineering information). There is also an extensive list derived from BS8888. Some examples are shown in Table 8.12.

| Symbol or abbreviation | Meaning | Symbol or abbreviation | Meaning |
|---|---|---|---|
| AF | Across flats | MIN | Minimum |
| CG | Centre of gravity | NTS | Not to scale |
| CH HD | Cheese head | REF | Reference |
| ∅ or DIA | Diameter | RH | Right hand |
| DRG | Drawing | SPEC | Specification |
| FIG | Figure | ⟶▷ | Taper |
| MAT | Material | THK | Thickness |
| MAX | Maximum | VOL | Volume |

**Table 8.12** Symbols and abbreviations

## Make the grade

This activity will help you in achieving the following grading criterion:

**P4** identify symbols and abbreviations used on given engineering documentation.

## Activity

**Use the information in Unit 2 and research BS8888 to complete Table 8.13 below.**

| Symbol or abbreviation | Meaning |
|---|---|
| ABS | Acrylonitrile butadiene styrene |
|  | Drawing number |
| —·—·—·— |  |
|  | Reference |
| Fe |  |
|  | Wire gauge |
| Mm |  |
|  | High speed steel |
| RSJ |  |
|  | Across flats |
| 1:5 |  |
|  | Diameter |
| BDMS |  |
|  | Material |
| PCB |  |
|  | Specification |
| ∅ |  |

**Table 8.13** Symbols and abbreviations

# Material documentation and coding systems

## Material documentation

If you buy a games console, MP3 player or mobile phone, you will find that you also receive documentation in the form of a booklet or leaflet. The documentation includes instructions, which are useful, but it may also include specifications and sometimes engineering drawings.

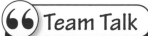

**66 Team Talk**

Aisha: **'What is a specification?'**
Steve: **'A specification gives detailed information on products that can be used by customers, suppliers and purchasers. The reason it is called a specification is because it details the specific requirements of a product. Specifications also include information on the quality standards used in the manufacture of a product.'** 99

**Activity**

List the things you will find in the documentation included with a games console or mobile phone. Compare your list with the actual documentation found with one of these products. Give yourself one mark for each correct answer.

Engineering drawings may be crucial if you purchase a product that needs to be assembled. A good example of this is a flat-pack bookcase where the documentation will include a step-by-step guide on how to assemble the product. However, a more detailed engineering drawing and specification will be included if you buy a part for a car engine and the material(s) in the product will be clearly labelled.

## Coding systems

Codes have been used throughout history to pass on information, using letters, numbers, symbols or even colours. In engineering, it is important to give materials and products their own codes so they can be quickly identified with regard to standards, material type, material colour, etc.

The quality of a material or product can be identified by certain standards. Two of the most common sets of standards are described below.

## British Standards Institution (BSI)

The British Standards Institution (BSI) currently has over 27,000 standards available. Some of the products that meet the criteria in the standards bear the BSI Kitemark. There is a basic coding system that is quite easy to follow. For example, the British Standard that gives a general statement for steel, concrete and composite bridges is BS5400. This standard is split into ten parts. The code for the first part would be BS5400-1:1988. The code works like this:

- 5400 is the number of the standard;
- 1 is the number that refers to the section of the standard (in this case, the '1' refers to a general statement; if it were '2', it would refer to a specification for loads on bridges);
- 1988 is the year that the standard came into effect.

## International Organisation for Standardisation (ISO)

The International Organisation for Standardisation (ISO) is similar to the BSI in that it sets standards for products, but the ISO does so for worldwide use. The ISO works with governments and professional bodies across the world to agree on standards. This is a difficult task when you take into account the number of different cultures and standards that exist. Meeting standards can be expensive, which means that a lot of the ISO recommendations become law (this can also result in safeguarding the public by insisting on standards such as health and safety). The coding system is very similar to the BSI. For example, the ISO standard that defines mechanical and physical properties for metric fasteners is ISO898. There are seven parts to this standard, so one of the parts is coded as ISO898-1.

Some suppliers and organisations have their own codes (sometimes letters and numbers, sometimes colour codes) so that materials can be identified quickly. Colour coding is commonly used on electronic components such as resistors. This is discussed in more detail in Unit 19 (Electronic circuit construction).

# Forms of supply

Ferrous and non-ferrous metals come in different shapes and sizes, known as 'forms of supply'. It is important for an engineer to know what forms of supply are available because they need to take into account strength, material removal, material waste and aesthetics when purchasing materials.

You will be familiar with the majority of the forms of supply available, but you may not know the correct terminology for them. Table 8.14 shows a range of forms of supply that are currently available.

## Activity

**In groups of two, use the information in Table 8.14 (Forms of supply) to create a card or board game.**

You may want to order a material that is an unusual shape – this can be done, but you would probably need to order one of the following forms of supply:

- Powders, grains, pellets and fluids – form of supply of plastics before they are used in a mould to form a shape.
- Castings and mouldings – metal or plastic is heated and set in a mould by cooling to form a casting. There are various techniques used to cast metals and to form plastics.
- Forgings – shaping materials with compressive forces (hammering). This can be made easier in some cases if the material is heated up first.

## Activity

**Make a bridge after designing it in Westpoint (see page 202). Choose your materials and say what form of supply they would be bought and delivered in.**

### Make the grade

The next activity will help you in achieving the following grading criterion:

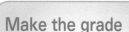 **M2** **select an appropriate form of supply for a given material requirement.**

# Size and surface finish

## Size

The size of a material needs to be taken into account at the design stage of a product. There are common sizes of materials readily available that can be used directly or

| Form of supply | Diagram | Form of supply | Diagram |
|---|---|---|---|
| Square bar | 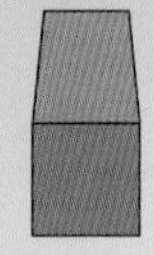 **Figure 8.20a** Square bar | Square tube | 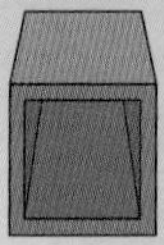 **Figure 8.20b** Square tube |
| Round bar | 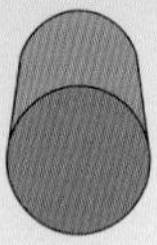 **Figure 8.20c** Round bar | Round tube | 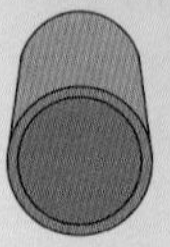 **Figure 8.20d** Round tube |
| Rectangular bar | 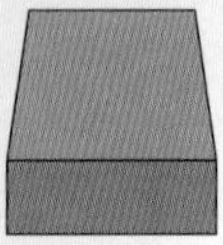 **Figure 8.20e** Rectangular bar | Rectangular tube | 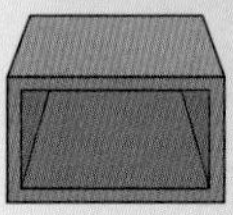 **Figure 8.20f** Rectangular tube |
| Hexagonal bar |  **Figure 8.20g** Hexagonal bar | Rod | 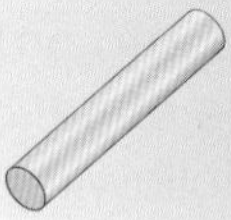 **Figure 8.20h** Rod |
| Plate |  **Figure 8.20i** Plate | Sheet | 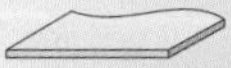 **Figure 8.20j** Sheet |
| Angle bar | 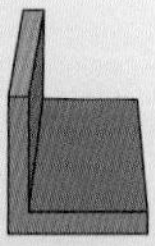 **Figure 8.20k** Angle bar | Channel | 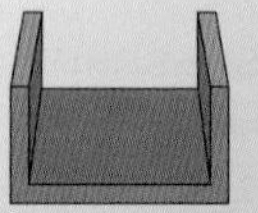 **Figure 8.20l** Channel |
| Beam | 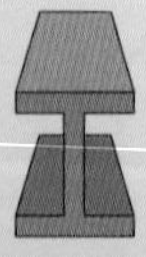 **Figure 8.20m** Beam | Wire | 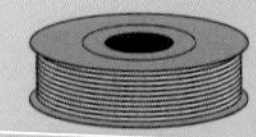 **Figure 8.20n** Wire |

**Table 8.14** Forms of supply

machined to size. If an unusual size is ordered, then the cost will increase.

Typical sizes for various forms of supplies of materials can be found in manufacturers' catalogues or on the internet.

The following examples are taken from www.metals4u. co.uk and show the sizes available along with the prices.

Tube/pipe – in this case, nylon is used but tube and pipe are available in a wide range of materials. OD stands for outside diameter of the tube and ID is the inside diameter. The lengths available are 100 mm, 250 mm, 500 mm, 1 metre, 2 metres and 3 metres. Prices in Table 8.15 are for lengths of 100 mm. Note: the price decreases when the pipe gets thinner.

| Size | Price |
|------|-------|
| 25 OD × 18 ID | £0.90 |
| 35 OD × 20 ID | £2.61 |
| 40 OD × 30 ID | £2.17 |
| 45 OD × 25 ID | £3.71 |
| 50 OD × 25 ID | £5.22 |
| 50 OD × 40 ID | £2.97 |
| 60 OD × 30 ID | £7.35 |
| 60 OD × 50 ID | £3.88 |
| 70 OD × 36 ID | £10.50 |
| 80 OD × 40 ID | £13.37 |
| 80 OD × 50 ID | £11.07 |
| 80 OD × 60 ID | £8.35 |

**Table 8.15** Tube/pipe sizes and prices

Solid diameter – when round bar is ordered, the only sizes required are the diameter and length. The following information is for stainless steel, which is typically available in 1-metre, 2-metre and 3-metre lengths. Note: diameters are in metric (mm) and imperial (inches) sizes, and only a selection of sizes is shown, as there is an extensive range available. Prices given in Table 8.16 are all for 1-metre lengths.

Plate/sheet – this can be typically ordered in the following dimensions:

- 500 mm × 500 mm;
- 1,000 mm × 1,000 mm;
- 2,000 mm × 1,000 mm.

Mild steel is available in this form of supply in a range of thicknesses, as shown in Table 8.17. Prices are based on 500 mm × 500 mm.

| Size | Price |
|---|---|
| 3 mm diameter | £4.37 |
| 4 mm diameter | £4.66 |
| 5 mm diameter | £5.03 |
| 6 mm diameter | £5.49 |
| $\frac{1}{4}$" diameter | £6.01 |
| 8 mm diameter | £4.40 |
| 10 mm diameter | £4.62 |
| 12 mm diameter | £4,89 |
| ½" diameter | £4.99 |
| 14 mm diameter | £12.11 |
| 15 mm diameter | £13.31 |
| $\frac{5}{8}$" diameter | £5.55 |

**Table 8.16** Round bar sizes and prices

| Thickness | Price |
|---|---|
| 1 mm | £16.10 |
| 1.5 mm | £16.60 |
| 2 mm | £24.00 |
| 3 mm | £30.00 |

**Table 8.17** Mild steel thicknesses and prices

Gauge size – the diameter or cross-sectional area of a wire can be referred to as the gauge (it is also referred to as 'standard wire gauge' (SWG)). This term is commonly used with electrical wires and guitar strings. The gauge of the wire gives an indication of the wire's properties, electrical or acoustic in this case. A table comparing wire gauges with diameters can be used for quick reference. An example of this is shown in Table 8.18.

## Activity

Use the examples in this chapter to collect a range of forms of supply of materials.

Calculate the heaviest etc (from earlier notes and other units) and test your results.

## Surface finish

The surface finish on a material usually depends on the desired functions of a product. Coating a material

| SWG | mm |
|---|---|
| 7/0 | 12.70 |
| 6/0 | 11.79 |
| 5/0 | 10.97 |
| 4/0 | 10.16 |
| 3/0 | 9.45 |
| 2/0 | 8.84 |
| 0 | 8.23 |
| 1 | 7.62 |
| 2 | 7.01 |
| 3 | 6.40 |
| 4 | 5.89 |
| 5 | 5.39 |

**Table 8.18** Standard wire gauges and diameters

in a certain way can enhance the product for the purchaser. There are different reasons for coating a material: to protect it from corrosion, to make it more durable, to enhance the aesthetics of a product, to improve the material's electrical or thermal insulation (or sometimes conductivity). We have already introduced the forms of supply and one of the main considerations to be taken when ordering a material is its surface finish.

**Figure 8.21** Gold-plated television leads

- Bright drawn – steel is produced by initially hot-rolling the raw material, then cold-working to improve the surface finish. This process enhances the mechanical properties as well as improving straightness and dimensional tolerances.

- Electroplating – metals such as nickel can be used to coat a material. Electroplating works by placing a cathode (the metal to be plated) into an electrolytic solution (a liquid that allows the flow of electricity) along with an anode (in this case, nickel). The anode starts to dissolve and coats itself onto the cathode as long as there is a current flowing through the circuit. Gold plating can be done in this way to enhance the appearance of a product or to change properties, such as electrical conductivity and durability in applications such as television leads.

- Hot-dip coating – this is a method of preventing corrosion on steel or iron. Zinc is normally used for this method and it works by dipping ferrous metals into molten zinc (approx 460 degrees Celsius) to apply a non-corrosive layer. This is commonly used on roofing, walling, car bodies, lamp posts and many other products used outdoors.

- Painting – one of the easiest ways to coat a material to prevent corrosion or enhance its aesthetics is by painting it; this can be done by brushing, spraying or dipping. This is a very easy and cheap method but it does have limitations. For example, the paint on a car can be easily scratched, as opposed to a metal coating. Paint also can react with water or chemicals, so it is important that the correct paint is used for a material.

- Plastic coating – there are various ways to coat a material in plastic. One of these methods is called powder coating. This is achieved by distributing a plastic powder over the material and 'curing' it in an oven; when the powder melts, it flows to form a coating (sometimes called a skin). The advantage of using this method instead of paint is that there are no solvents used, so it is more environmentally friendly. Paints use solvents to keep the solid parts in liquid form until the solvents evaporate to leave a solid coating.

**Make the grade**

The next activity will help you in achieving the following grading criterion:

**P5**  identify the forms of supply available for a given engineering material.

## Activity

**Identify six forms of supply available for mild steel. Sketch them and give an application of each form of supply, using Table 8.19 below. An example is provided to get you started.**

| Form of supply | Sketch | Application |
|---|---|---|
| Round tube | | Bicycles or pipes |
| | | |
| | | |
| | | |
| | | |

**Table 8.19** Forms of supply for mild steel

## Make the grade

The next activity will help you in achieving the following grading criterion:

**D1**  establish that a material has the required properties for a given application.

## Activity

Choose a suitable material and confirm that it has the required properties for a car body. Use the results from material tests to support your conclusion and take the following into consideration:

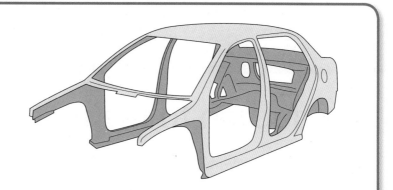

**Figure 8.22** Car body

- cost;
- availability;
- machinability;
- strength;
- toughness;
- corrosion properties.

 **Grading criteria recap**

To achieve a pass grade you must be able to:

**P1** describe the properties that are used to define the behaviour of common engineering materials;

**P2** review the properties of a given ferrous metal and a given non-ferrous metal;

**P3** review the properties of a given organic material, a given thermoplastic, a given thermosetting polymer and a given smart material;

**P4** identify symbols and abbreviations used on given engineering documentation;

**P5** identify the forms of supply available for a given engineering material.

To achieve a merit grade you must be able to:

**M1** explain the choice of material for a given engineering component based on the material's properties;

**M2** select an appropriate form of supply for a given material requirement.

To achieve a distinction grade you must be able to:

**D1** establish that a material has the required properties for a given application.

# Unit 10
## Using computer-aided drawing techniques in engineering

# Introduction to the unit

Computer-aided drawing (CAD) is generally accepted as the first-choice method for producing engineering drawings. In this unit, you will be introduced to CAD systems, discover how they are used and learn what makes them more efficient than manual drafting. CAD is a part of what is now referred to as computer-integrated manufacturing (CIM), which also encompasses computer-aided manufacturing (CAM) and computer-aided engineering (CAE). The purchase of a CAD system for an engineering design office will not in itself produce designs at the click of a mouse button, neither is the draughtsman obsolete because of the emergence of CAD systems. Engineering design still requires decisions to be made at all stages from conception to production. The computer is merely an aid to the person using it. A CAD system performs as directed and with accuracy and speed.

## Learning Outcomes

By the end of this unit you should:

- be able to start up and close down hardware and software in order to perform CAD activities;
- be able to produce CAD drawings;
- be able to modify engineering drawings using CAD commands;
- be able to store and retrieve engineering drawings for printing/plotting.

# Grading criteria

| To achieve a pass grade you must be able to: | To achieve a merit grade you must be able to: | To achieve a distinction grade you must be able to: |
|---|---|---|
| P1 start up a CAD system, produce and save a standard drawing template and close down CAD hardware and software in the approved manner | M1 identify and describe four methods used to overcome problems when starting up and closing down CAD hardware and software | D1 justify the use of CAD for the production of a range of drawing types |
| P2 produce a CAD drawing using an orthographic projection method | M2 describe the drawing commands used across the range of drawing types | D2 demonstrate an ability to produce detailed and accurate drawings independently and within agreed timescales |
| P3 produce a CAD drawing using an isometric projection method | M3 describe the methods used to create relevant folder and file names and maintain directories to aid efficient recovery of data | |
| P4 produce a circuit diagram using CAD | | |
| P5 use CAD commands to modify a given orthographic and isometric drawing | | |
| P6 use CAD commands to modify two different given circuit diagram types | | |

| | | |
|---|---|---|
| **P7** set up an electronic folder for the storage and retrieval of information | | |
| **P8** store, retrieve and print/plot seven CAD-generated or modified drawings | | |

# Learning Outcome 1.  Be able to start up and close down hardware and software in order to perform CAD activities

## Starting up, using and closing down a CAD system

**P1**  **M1**

Many households in the United Kingdom own at least one computer and some may have more. In the modern home, we quite often take the personal computer (PC) for granted. Even without the benefit of specific training on a computer, you can no doubt easily manage to start up your system, use a variety of software packages and then close down your system correctly. You may only use basic software applications, such as email, web browsers, photo editing and games; however, while doing so, you are using a series of menu commands and functions that are similar to those used in most CAD packages. A modern home computer is also likely to be more powerful than a top-of-the-range CAD computer of a decade or so ago.

### Computer hardware basics

A typical PC system is shown in Figure 10.1. The physical equipment components of the system are known collectively as the hardware. Programs and data used on the computer are known as software.

We need to input data to the computer and, for this, we use input devices, such as the mouse and keyboard. These data are then processed and stored by components

housed within the tower case or chassis. Devices such as the LCD screen or a printer are used to output useable data or images. We can also output data and information onto removable media, such as discs or memory cards. Devices that are attached to the main computer are usually known as peripheral devices. These can be external peripheral devices (e.g. the mouse) or internal peripheral devices (e.g. the DVD-ROM drive). Modern computers have plug-and-play (PnP) capability, which enables them to connect easily to peripheral devices.

**Figure 10.1** Computer hardware

**Team Talk**

Aisha: **'So what does a CAD system look like?'**
Steve: **'It could look exactly like the one shown in Figure 10.1! Most modern PCs are powerful enough to run basic CAD software.'**

## Computer-aided drawing hardware

Home computers are suitable for basic CAD, but for engineering and design professionals, a more powerful computer with enhanced capabilities is often required. Typically, a purpose-built CAD system will have a very fast processor and a great deal more memory than your home computer. Input devices, such as scanners, graphics tablets and digitisers, are also used as well as the mouse. If you are dealing with very complex engineering drawings and you need to view a lot of information on screen, then a dual-monitor set-up is sometimes used. Engineering drawings are often much larger than the common A4-sized print used at home; therefore, large printers or plotters are required. In Figure 10.2, a graphics tablet is being used as an input device instead of a mouse.

**Figure 10.2** Typical CAD workstation

There is a vast array of external peripheral hardware for computers. Ideally, any devices that you need to connect to your computer system should have PnP capability (FireWire® and USB connections are PnP). Once the PnP device is connected, the computer recognises the hardware, loads the correct software drivers and the device can be used straight away. This makes installing devices such as scanners, digitisers, printers and plotters a simple task. Once you have connected all the peripheral devices and the computer is switched on, you are ready to start the CAD package.

## Operating systems

Computers require an operating system (sometimes abbreviated to OS) in order for us to interface with the computer hardware. It could be said that the OS runs the computer. Operating systems such as Windows Vista® and XP® are commonly used; however, others are available (e.g. MAC OS X®, Linux® and Unix®).

## Applications

A software application is a computer program that is designed to help the user perform a particular task. Typical software application packages are used for spreadsheets, databases, word processing and photo editing. A CAD package, therefore, is also a software application.
There are many CAD packages available and they vary considerably in functionality and price. AutoCAD® is a mainstream CAD package used in engineering. It is used for general engineering drawings in two dimensions (2D), such as floor plans, schematic diagrams and line diagrams (see Figure 10.3). For more complex engineering and

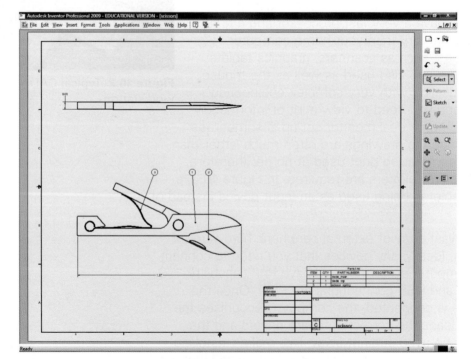

**Figure 10.3** A 2D drawing of a pair of cutters

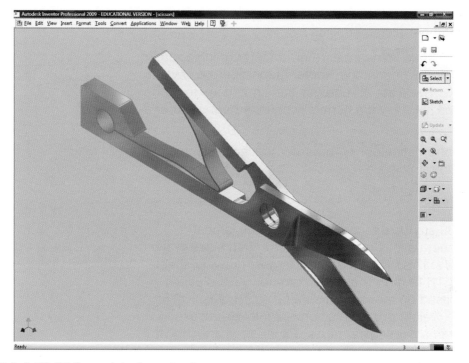

**Figure 10.4** A 3D CAD model of a pair of cutters

design tasks, CAD packages that can produce drawings in three dimensions (3D) are required (see Figure 10.4). These are also known as solid modelling or parametric drawing packages. Google SketchUp™ is a popular drawing package that was designed to be easier to use than other 3D design packages. The advantage of 3D designs over 2D is the ability to view a representation of the finished design (sometimes called the model) on screen. It can also be changed and edited on screen. With so many CAD packages available, one important consideration is the compatibility of the file types between each different package.

## How do CAD applications work?

CAD programs use a mathematical technique known as vector graphics to produce an image on screen. Vectors are used to describe an object's specific size, position and geometry.

When you view photographic images on screen, these are represented by a different technique known as raster graphics. For example, a bitmap is a grid of pixels or points of colour that can be viewed on screen or paper. It is generally characterised by its width and height in pixels and the number of bits per pixel.

Printing industries sometimes refer to raster graphics as continuous tone, or contone. Vector graphics are known as line work. There are advantages and limitations to both technologies and they are generally complementary to one another.

For more infromation on vectors refer to Unit 4 Electrical and Mechanical Science.

## Paper sizes

The BS8888 standard recommends that the largest drawing sheet used is one square metre with sides in the ratio of 1: $\sqrt{2}$. This would make the largest drawing sheet 841 mm × 1,149 mm. The more commonly used paper sizes are based on the longest side being progressively halved. The weight of paper is conveniently expressed in units of 'grams per square metre'. Paper sizes are shown in Table 10.1.

### Activity

**Measure a sheet of A4 paper. Determine (or prove) the ratio of 1: $\sqrt{2}$.**

| Designation | Size (millimetres) | Area |
| --- | --- | --- |
| A0 | 841 × 1,189 | 1 m² |
| A1 | 594 × 841 | 5,000 cm² |
| A2 | 420 × 594 | 2,500 cm² |
| A3 | 297 × 420 | 1,250 cm² |
| A4 | 210 × 297 | 625 cm² |

**Table 10.1** Standard ISO paper sizes

Title blocks are generally placed in the bottom right corner of drawings. They contain items of basic information required by the user of the drawing or by a drawing office. Typical information contained in the title block would be:

● drawing title;
● name of firm or company logo;
● drawing number;
● component name;
● drawing scale;

- units of measurement;
- projection method;
- date of drawing;
- the name of the drawing author;
- cross references or associated drawing numbers.

Other information provided is dependent on the sector of industry concerned. For more information on this see Unit 2 Interpreting and Using Engineering Information.

## Activity

Complete Table 10.2 below, listing examples of computer hardware and software in each column. Indicate whether they are input or output devices, operating systems or applications. The table contains an example of each to help you get started.

| Hardware | | Software | |
|---|---|---|---|
| **Input device** | **Output device** | **Operating system** | **Application** |
| Mouse | Printer | Microsoft Windows® XP | Internet Explorer® |
| | | | |
| | | | |
| | | | |

**Table 10.2** Computer hardware and software

## Starting up and closing down CAD hardware and software

Turning on a CAD system is very much the same as turning on a home computer. As long as they are connected correctly, any peripheral devices will power up when the computer is booted up. To open or start the CAD program, you can double-click the relevant desktop icon or select the program from the Start menu. Figures 10.5 and 10.6 illustrate how this is done using AutoCAD® 2010.

**Figure 10.5** AutoCAD® 2010 desktop icon

The main window that appears when you open a CAD package is called the user interface (UI). The appearance and functionality of each UI will differ depending on the CAD software used. The UI for AutoCAD® 2010 is shown in Figure 10.7, with a CAD drawing already opened. Each command within the CAD program can be accessed from dropdown menus, by clicking on icons or by typing into a command line.

**Figure 10.6** Starting a CAD program from the Windows® Start menu

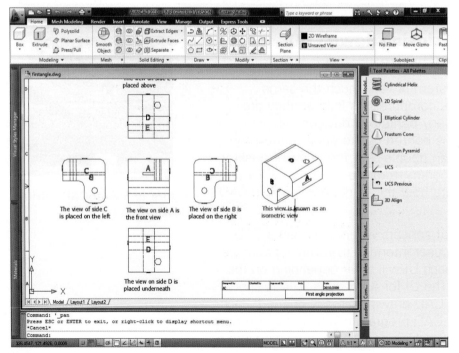

**Figure 10.7** The AutoCAD® 2010 user interface

The UI can appear complex as there are many different icons and commands available. The appearance of the UI can be customised by removing or hiding icons and functions. Figure 10.8 shows the dropdown application menu and the quick-access toolbar.

Once the CAD program is running, you can start a new drawing from scratch or load a template file. Using template files ensures consistency in the drawings you create by using standard settings and styles. Template files have certain parameters preset, allowing you to work more efficiently. Some CAD packages may also have a wizard-type help system to guide you through the template set-up process.

Drawing units can be set to metric or imperial and the precision to which the dimensions and tolerances are set can also be selected. Each drawing unit represents a unit of measurement in the system you choose. For example, if you select metric measurements, then one drawing unit can represent one millimetre, one centimetre or one metre, depending on the one you select. Thus, your drawing is created at actual size depending on the convention used, although it appears scaled down on screen. A mechanical engineer might want to use decimal units and an architect might prefer to use feet and inches.

Other parameters, such as paper size, logos, toolbars, layers, line types, text styles, title and border styles, can also be preset in a template. To create a new template,

**Figure 10.8** A dropdown application menu and quick-access toolbar

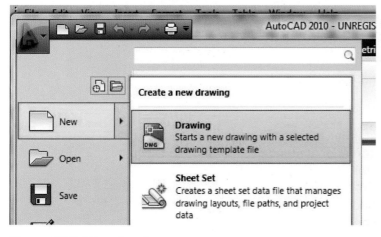

**Figure 10.9** New drawing menu

you need to set up your drawing with the parameters you require and save the drawing with a template file extension. AutoCAD® 2010 uses the .dwt file type extension for templates. Figure 10.9 illustrates the use of the New command from the application menu.

If you open a template file, you are still creating a new drawing file based on that template.

# Learning Outcome 2. Be able to produce CAD drawings

## Producing CAD drawings

P2   P3   P4   M2   D1

D2

### Drawing standards

Engineering drawings are used to provide technical information. Engineering components and systems could be produced from them, so they need to be accurate and easy for the user to understand. To maintain accuracy and ensure they are fit for purpose, drawings have to be prepared using common standards. The standards that need to be taken into account are called ISO and British Standards. In general, an engineering drawing will be to a good standard if the following points have been observed:

- Line types used are of uniform thickness and density.
- Avoid the use of unnecessary artwork, such as shading, printing or artistry.
- Include only enough information to ensure accurate and clear communication.
- Use standard symbols and abbreviations.
- Ensure the drawing is correctly dimensioned.

### ISO standards

ISO standards are international standards recognised in industry and engineering. The international standards body is the International Organisation for Standardisation, but it is more commonly referred to as the ISO standards. The organisation itself is comprised of national standards organisations from around the world.

The standards used depend on the country where the drawing is produced. BS8888 is the standard used in the United Kingdom for engineering drawing and technical product specifications. It covers any drawings that define shape, size and form, whether produced by manual drafting methods or by CAD.

## British Standards

The British Standards Institute was the world's first national standards organisation. BSI has published around 20,000 standards and every year almost 2,000 new and revised standards are issued. Standards need to be renewed and revised regularly to take into account developments in new technology, materials and processes.

BS8888:2008 is an important standard and refers to over 130 ISO and EN ISO standards, which must be complied with. It specifies not only the use of general engineering drawing principles, such as line types, dimensioning and borders, but also all aspects of technical product specifications (including 3D CAD models). The standard is normally reviewed every two years.

## Standard symbols and abbreviations

Engineering drawings need to transmit technical information and be easily understood independently from any language. Therefore, an engineering drawing should be understood by everybody, regardless of their native language. To achieve this, standard symbols and abbreviations are used. Symbols are used to reduce sources of confusion by using a standard graphical representation for a part or component. Abbreviations are used to prevent drawings becoming cluttered and confusing. Further information on this is available in Unit 2.

### Symbols

A range of electrical and electronic components is represented by the symbols shown in Figure 10.10.

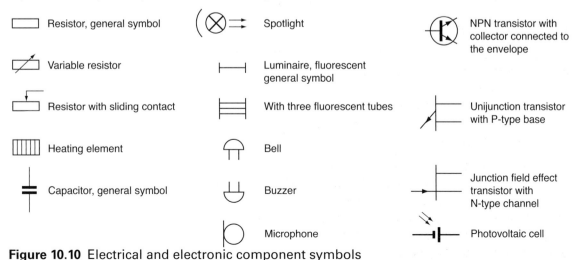

**Figure 10.10** Electrical and electronic component symbols

Fluid-power equipment (pneumatics and hydraulics, for example) can be represented by the standard symbols shown in Figures 10.11 and 10.12.

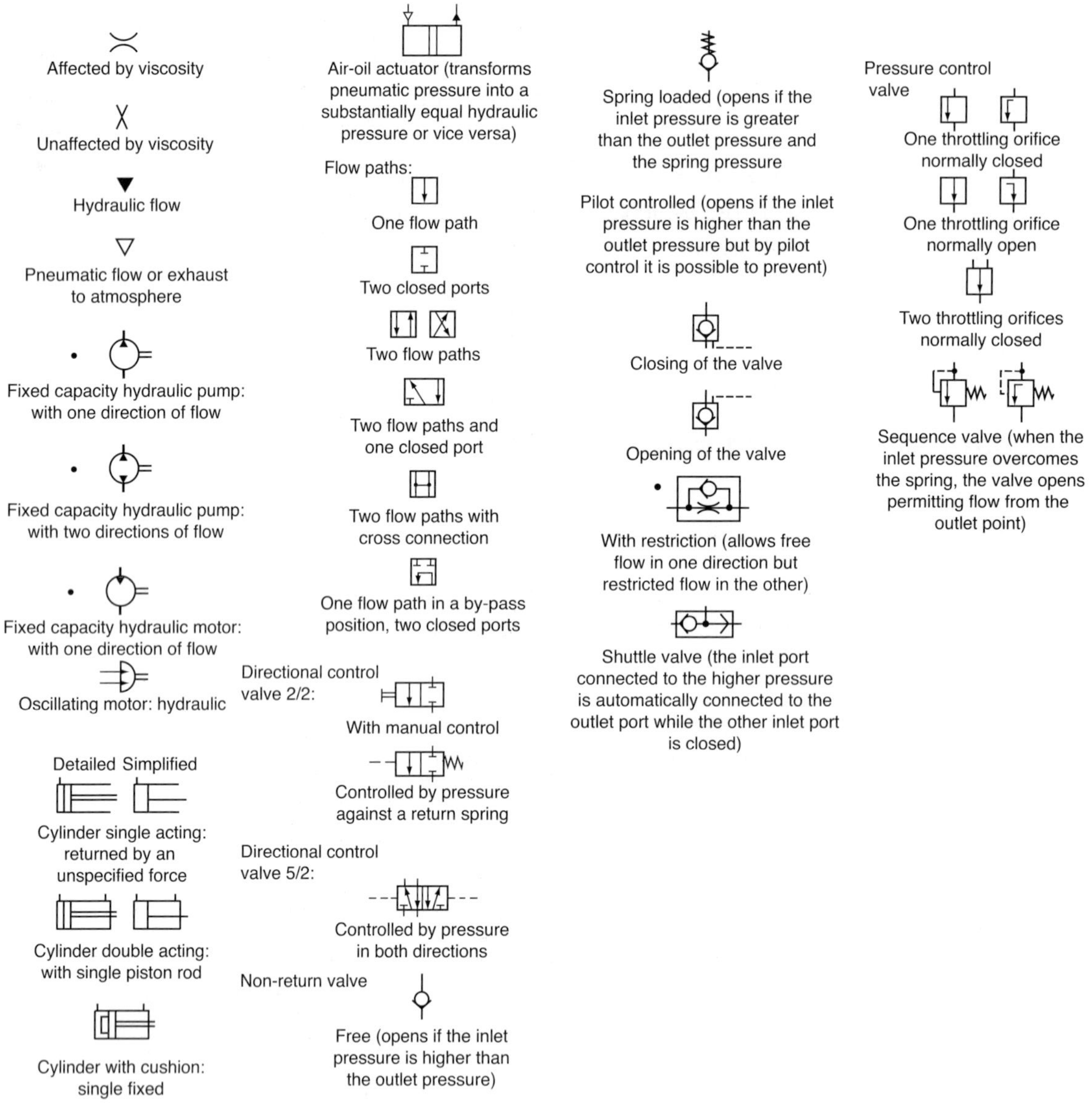

Affected by viscosity

Unaffected by viscosity

Hydraulic flow

Pneumatic flow or exhaust to atmosphere

Fixed capacity hydraulic pump: with one direction of flow

Fixed capacity hydraulic pump: with two directions of flow

Fixed capacity hydraulic motor: with one direction of flow

Oscillating motor: hydraulic

Detailed  Simplified

Cylinder single acting: returned by an unspecified force

Cylinder double acting: with single piston rod

Cylinder with cushion: single fixed

Air-oil actuator (transforms pneumatic pressure into a substantially equal hydraulic pressure or vice versa)

Flow paths:

One flow path

Two closed ports

Two flow paths

Two flow paths and one closed port

Two flow paths with cross connection

One flow path in a by-pass position, two closed ports

Directional control valve 2/2:

With manual control

Controlled by pressure against a return spring

Directional control valve 5/2:

Controlled by pressure in both directions

Non-return valve

Free (opens if the inlet pressure is higher than the outlet pressure)

Spring loaded (opens if the inlet pressure is greater than the outlet pressure and the spring pressure

Pilot controlled (opens if the inlet pressure is higher than the outlet pressure but by pilot control it is possible to prevent)

Closing of the valve

Opening of the valve

With restriction (allows free flow in one direction but restricted flow in the other)

Shuttle valve (the inlet port connected to the higher pressure is automatically connected to the outlet port while the other inlet port is closed)

Pressure control valve

One throttling orifice normally closed

One throttling orifice normally open

Two throttling orifices normally closed

Sequence valve (when the inlet pressure overcomes the spring, the valve opens permitting flow from the outlet point)

**Figure 10.11** Fluid-power symbols 1

Throttle valve: simplified symbol

Braking valve

Flow control valve with fixed output

Flow dividing valve with fixed output

Pressure source

Electric motor

Heat engine

Flow line: working line, return line and feed line

Flow line: pilot control line

Flow line: drain or bleed line

Flow line: flexible pipe

Pipeline junction

Crossed pipelines (not connected)

Air bleed

Power take-off: plugged

Power take-off: with take-off line

Connected, with mechanically opened non-return valves

Uncoupled, with open end

Uncoupled, closed by free non-return valve

Rotary connection: one way

Rotary connection: three way

Reservoir open to atmosphere

With inlet pipe above fluid level

With inlet pipe below fluid level

With a header line

Pressurized reservoir

Accumulators: The fluid is maintained under pressure by a spring, weight or compressed gas

Filter or strainer

Temperature controller

Cooler with representation of the flow lines of the coolant

Heater (arrows indicate the introduction of heat)

Rotating shaft: in one direction

Rotating shaft: in either direction

Detent (device for maintaining a given position)

Locking device

Over-centre device (prevents stopping in a dead center position)

Pivoting devices: simple

Pivoting devices: with traversing lever

Pivoting devices: with fixed fulcrum

Muscular control: general symbol

Push-button

Lever

Pedal

Mechanical control: Plunger or tracer

Spring

Roller

Roller operating in one direction only

Electrical control: by solenoid (one winding)

Electrical control: by electric motor

Direct acting control: by application of pressure

Direct acting control: by release of pressure

Combined control: by solenoid and pilot directional valve (pilot directional valve is actuated by the solenoid)

Pressure measurement: pressure gauge

Pressure electric switch

**Figure 10.12** Fluid-power symbols 2

## Activity

Using the information provided in Figures 10.11 and 10.12, identify the components highlighted in the fluid-power circuit in Figure 10.13.

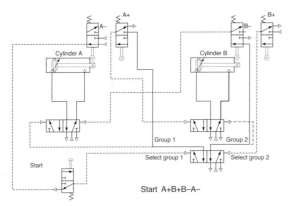

**Figure 10.13** Fluid-power circuit diagram

## Activity

1.  **What standard references all aspects of technical product documentation?**
2.  **What are the main factors that need to be observed to ensure an engineering drawing is produced to a good standard?**

## Abbreviations

In order to shorten drawing notes, common engineering terms and phrases are abbreviated to a standard format. Some examples are given in Table 10.3.

| Term | Abbreviation |
|------|--------------|
| Across the flats | A/F |
| Centres | CRS |
| Diameter | DIA |
| Left hand | LH |
| Pitch circle diameter | PCD |

**Table 10.3** Common engineering abbreviations

## Projection methods

When you view an engineering component drawing, it is useful if you are able to see the object drawn from all sides. Orthographic projection is a method of displaying the object as though it is being viewed from the front, side, top and bottom. The top view is usually referred to as the plan view.

## Activity

**What other abbreviations are used on engineering drawings? Make a list of further examples.**

There are two ways of drawing in orthographic projection; these are known as first angle and third angle. They differ only in the position in which each view

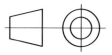

First angle projection

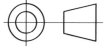

Third angle projection

**Figure 10.14** Orthographic projection symbols

is placed on the drawing. First-angle projection is used mainly in Europe and third-angle projection is used mainly in the United States.

In this age of modern telecommunications, it is possible for CAD drawing files to be sent around the globe electronically. Therefore, both projection systems are regularly in use; it is not acceptable, however, to use both projection systems on the same drawing.

Special symbols are used on engineering drawings to indicate which projection symbol has been used. The symbols for first- and third-angle projection are shown in Figure 10.14.

A shaped solid block is shown in Figure 10.15.

Each face of the block has a letter to represent the view seen directly on each face.

- Viewing the block in direction A is the front view.
- Viewing the block in direction B is the view from the left.
- Viewing the block in direction C is the view from the right.
- Viewing the block in direction D is the view from above (plan view).
- Viewing the block in direction E is the view from below.

**Figure 10.15** A 3D solid model

## First-angle projection

To illustrate how the block would appear when drawn in first-angle projection, Figure 10.16 shows each of the views. An isometric view is sometimes included to illustrate the object in 3D. The layout of the views in first-angle projection in relation to the front view is as follows:

- The view from B is placed on the right.
- The view from C is placed on the left.
- The view from D is placed below.
- The view from E is placed above.

## Third-angle projection

The difference between first- and third-angle projection is in the layout of the views in relation to the front view.

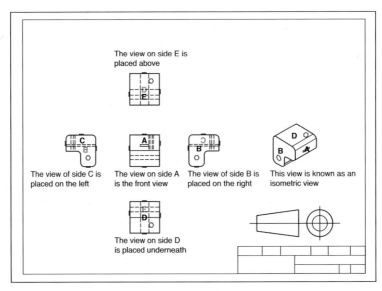

**Figure 10.16** First-angle orthographic projection (dashed lines indicate hidden lines)

Figure 10.17 shows the third-angle projection views now positioned as follows:

- View A is the front view.
- View B from the left is placed on the left.
- View C from the right is placed on the right.
- View D from above is placed above.
- View E from below is placed below.

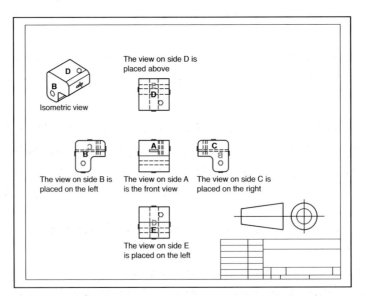

**Figure 10.17** Third-angle orthographic projection (dashed lines indicate hidden lines)

Traditionally, front views are known as front elevations, side views are often termed side or end elevations and plan views are used to describe views from above or beneath. To portray a simple engineering component or object, it is not always necessary to include all five views on the drawing. The CAD designer must decide the minimum number of views required to illustrate the maximum visual information.

---

## Make the grade

This activity will help you in achieving the following grading criteria:

**P1** start up a CAD system, produce and save a standard drawing template and close down CAD hardware and software in the approved manner;

**P2** produce a CAD drawing using an orthographic projection method;

**P7** set up an electronic folder for the storage and retrieval of information;

**D2** (in part) – demonstrate an ability to produce detailed and accurate drawings independently and within agreed timescales.

---

## Activity

Produce a CAD drawing using an orthographic projection method.

1. Start up a CAD system and set up a drawing template.
2. Produce a CAD drawing using an orthographic projection method. Start with basic exercises such as drawing standard symbols. Your tutor or supervisor may provide examples.
3. Save the template and orthographic projection drawing in an electronic folder.
4. Complete all work in a reasonable time period and to agreed and appropriate standards.

This task is a practical activity but you will need to keep a record of the steps you used to complete it. You can use screen dumps by pressing the print screen (PrtScn) key and pasting the captured images into a Word document. Make a detailed list of the main steps you took to complete this activity. For example, how you completed the following:

- switching the computer on and starting the CAD hardware and software;
- procedure to log in to a network;
- producing the template and orthographic drawing;
- saving the template and orthographic drawing to an electronic folder;
- closing the CAD software and shutting down the CAD hardware.

Obtain a witness statement from your instructor to show you have completed this activity competently and achieved the criteria.

# Single-part drawings

Single-part drawings are used to convey enough detailed information for a part or component to be manufactured without having to refer to other sources. The drawing shows all the sizes and locations of any features on the part or component. The full specification for the part would be included with details of dimensions and tolerances, type of material used and surface finish. Figure 10.18 illustrates a typical single-part drawing.

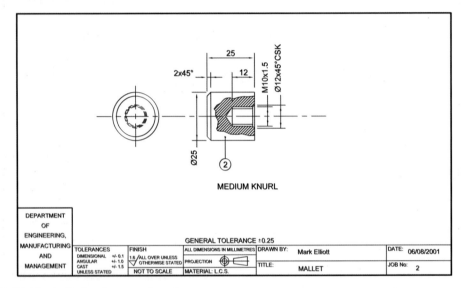

**Figure 10.18** Typical single-part drawings

Several small part details can be grouped together on the same drawing sheet when they each relate to the same assembly. This is useful when all the parts are to be manufactured in the same department in a factory.

# Assembly drawings

An assembly drawing is used to show how a complete product looks when all its separate component parts are assembled together. Figure 10.19 is an example of an assembly drawing.

Each component can be numbered using 'balloons' containing the part number. These are connected to the part by leader lines. Parts lists can be included and these give information on single-part numbers, quantities required and materials. Other information given on assembly drawings could include overall dimension sizes, shipping weights and operating instructions.

Exploded-assembly drawings are used to show the assembly with all the component parts slightly separated. This is useful for illustrating the intended assembly or disassembly sequence of each of the parts. Car manuals and servicing instructions for equipment and appliances generally include exploded-assembly drawings.

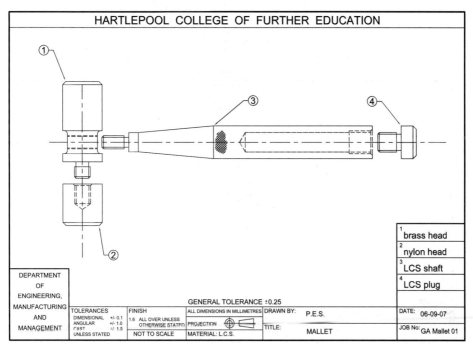

**Figure 10.19** Final-assembly drawing

# Isometric drawings

An isometric drawing is used when a 3D pictorial view of an object is required. Isometric views require only vertical lines and lines drawn at 30 degrees to the horizontal. A graphical representation of this is shown in Figure 10.20.

When using manual drawing instruments to produce an isometric drawing, careful measurement of angles and line lengths is required. Complex hand-drawn isometric views would require a great deal of time to complete. Even until fairly recently, drawing in isometric view when using CAD was a fairly complex task. Advances in CAD software have, however, simplified this task even further. For example, Autodesk Inventor® can generate multiple isometric views of the base model automatically from the projected view command (Figure 10.21).

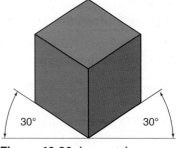

**Figure 10.20** Isometric convention

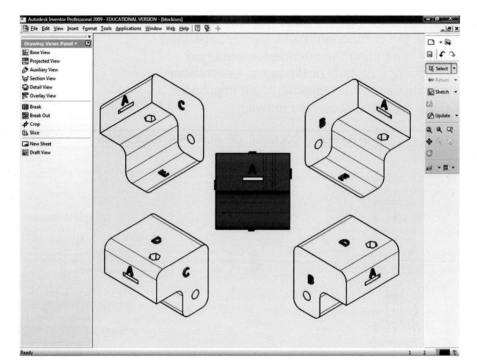

**Figure 10.21** Isometric views in Autodesk Inventor®

## Make the grade

The next activity will help you in achieving the following grading criteria:

**P3** produce a CAD drawing using an isometric projection method;

**P5** use CAD commands to modify a given orthographic and isometric drawing;

**P8** (in part) – store, retrieve and print/plot seven CAD-generated or modified drawings;

**M1** identify and describe four methods used to overcome problems when starting up and closing down CAD hardware and software;

**M3** describe the methods used to create relevant folder and file names and maintain directories to aid efficient recovery of data;

**D2** (in part) – demonstrate an ability to produce detailed and accurate drawings independently and within agreed timescales.

## Activity

1. Produce a CAD drawing using an isometric projection method.
2. Modify given CAD orthographic and isometric drawings and save them to an electronic folder.
3. Print/plot out seven CAD-generated or modified drawings and include them with the report.
4. Describe how you overcame any difficulties encountered during the start-up and shutdown of the CAD hardware and software.
5. Describe how you created the electronic folders and saved the CAD drawings.
6. Complete all work in a reasonable time period and to agreed and appropriate standards.

This task is a practical activity but you will need to keep a record of the steps you used to complete it. You can use screen dumps by pressing the print screen (PrtScn) key and pasting the captured images into a Word document. Produce a report describing the methods used to complete this activity.

Obtain a witness statement from your tutor or supervisor to show you have completed this activity competently and achieved the criteria.

## Circuit diagrams

A circuit diagram is a simplified diagrammatical representation of a system loop. We immediately think of an electrical circuit when considering circuit diagrams. There are, however, many other systems that can be represented by a circuit diagram. For example:

- hydraulic systems;
- pneumatic circuits;
- gas-supply piping;
- electronic circuits;
- water supply and drainage piping;
- central-heating piping;
- heating, ventilation and air-conditioning (HVAC) ductwork.

In all these types of diagrams, the component parts should be arranged neatly, with the circuit lines drawn horizontally or vertically. It is good practice to arrange the circuits so that the sequence of events can be read from left to right or top to bottom. It may even be a combination of both. The true shape and exact position of each circuit component are not reflected by the standardised symbols in the circuit layout. Figure 10.22 shows an example of a hydraulic circuit and Figure 10.23 shows an example of an electronic circuit.

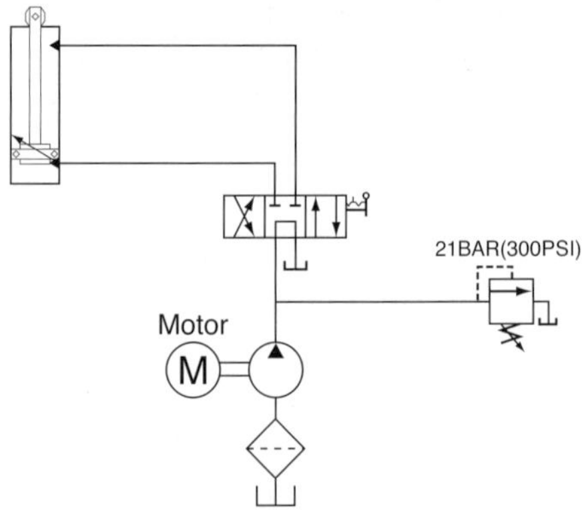

# Drawing commands

CAD hardware and software can be used to produce engineering drawings ranging from the most basic components to complex solid models. The novice engineering designer first needs to learn how to set up the CAD drawing environment. The next step is to master the basic drawing commands.

## Coordinate systems

When you enter a command in CAD (such as drawing a line), the software requires a location on the screen to start from and a location on the screen to end at. You can use the mouse to select a point on screen or you can specify a location by typing in a set of coordinates. You can enter two-dimensional coordinates as either Cartesian (X, Y) or polar coordinates, which use a distance and an angle to locate a point.

A Cartesian coordinate system has three axes, X, Y, and Z. When you enter coordinate values, you indicate a point's distance (in units) and its direction (+ or –) along the X, Y, and Z axes, relative to the coordinate system origin (0, 0, 0).

In 2D, you specify points on the XY plane, also called the 'workplane'. The workplane is similar to a flat sheet of graph paper. The X value of a Cartesian coordinate specifies horizontal distance and the Y value specifies vertical distance. The origin point (0, 0) indicates where the two axes intersect.

Polar coordinates use a distance and an angle to locate a point. With both Cartesian and polar coordinates, you can enter absolute coordinates based on the origin (0, 0) or relative coordinates based on the last point specified.

Another method of entering a relative coordinate is by moving the cursor to specify a direction and then entering a distance directly. This method is called direct distance entry.

You can enter coordinates in scientific, decimal, engineering, architectural or fractional notation. You can enter angles in grads, radians and surveyor's units, or degrees, minutes, and seconds. The Units command controls unit format.

**Figure 10.22** Hydraulic circuit

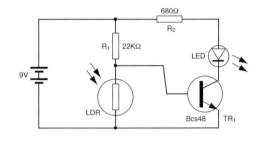

**Figure 10.23** Electronic circuit

You can find further information in Unit 3 Mathematics for Engineering Technicians, about the theory behind coordinates and other graphical methods.

## Line types

Most CAD packages come with a built-in line type library. The line type is not generally considered when 3D modelling as these are more concerned with solids and surfaces. However, it is important that the correct line types are used for 2D drafting. Examples of line types are shown in Figure 10.24.

There are usually only two thicknesses of line required in both CAD and manual drafting. A wide line and a narrow line in the ratio of 2:1 give the best results. Line thicknesses of 0.7 mm and 0.35 mm are widely used.

## Drawing aids

One main factor associated with using CAD as opposed to manual drafting methods is the accuracy and repeatability CAD offers. There are many built-in features within CAD packages that assist the user in producing highly accurate drawings. Two of these are the Grid and Snap functions.

| Linetype | Application |
|---|---|
| Continuous wide line | Visible edges and outlines |
| Continuous narrow line | 1 Dimension extension and projection lines<br>2 Hatching lines for cross sections<br>3 Leader and reference lines<br>4 Outlines of revolved sections<br>5 Imaginary lines of intersection<br>6 Short centre lines<br>7 Diagonals indicating flat surfaces<br>8 Bending lines<br>9 Indication of repetitive features |
| Continuous narrow irregular line | Limits of partial views or sections provided the line is not an axis |
| Dashed narrow line | Hidden outlines and edges |
| Long dashed dotted narrow line | 1 Centre lines<br>2 Lines of symmetry<br>3 Pitch circle for gears<br>4 Pitch circle for holes |
| Long dashed dotted wide line | Surfaces which have to meet special requirements |
| Long dashed dotted narrow line with wide lines at ends and at changes to indicate cutting planes | Note BS EN ISO 128–24 shows a long dashed dotted wide line for this application |
| Long dashed double dotted narrow line | 1 Preformed outlines<br>2 Adjacent parts<br>3 Extreme positions of moveable parts<br>4 Initial outlines prior to forming<br>5 Outline of finished parts<br>6 Projected tolerance zones |
| Continuous straight narrow line with zig zags | Limits of partial or interrupted views;<br>Suitable for CAD drawings provided the line is not an axis |

**Figure 10.24** Line types

These features set up a framework that can be used while drawing.

Grid is a rectangular pattern of dots that extends over the area you specify as the grid limits. Using the grid is similar to placing a sheet of grid paper under a drawing. The grid helps you align and visualise the distance between objects. The grid does not appear in the plotted drawing.

## Activity

1. To open AutoCAD®, click Start menu (Windows®) – (All) Programs – Autodesk – AutoCAD 2010 – AutoCAD 2010 – English. The AutoCAD window opens with an empty drawing file named Drawing1.dwg.
2. Check the status bar to make sure you are in the 2D Drafting and Annotation workspace.
3. On the status bar, click the Grid Display button to turn it on.

Snap restricts the movement of the crosshairs cursor to an interval that you define. When Snap mode is on, the cursor adheres or 'snaps' to an invisible grid. Snap mode is useful for specifying precise points with the cursor.

## Activity

The following activities will guide you through setting up the drawing environment and producing basic shapes with CAD hardware and software. Examples of a basic set-up procedure and some of the main drawing commands using AutoCAD® 2010 are shown.

Set up the drawing environment:

1. To open AutoCAD, click Start menu (Windows®) – (All) Programs – Autodesk – AutoCAD 2010 – AutoCAD 2010 – English.
2. The AutoCAD window opens with an empty drawing file named Drawing1.dwg.
3. Check the status bar to make sure you are in the 2D Drafting and Annotation workspace.
4. You must now specify some drafting settings. Right-click the Object Snap icon in the status bar at the bottom of the window, and select Settings.
5. The Drafting Settings dialog box is displayed (Figure 10.25). In the Object Snap tab, make sure the Object Snap On option is selected. Also, select Endpoint, Midpoint, and Centre for the Object Snap modes. Make sure the other options are not selected. Note the numeric shortcut key (F3) for switching Object Snap on. Select OK.

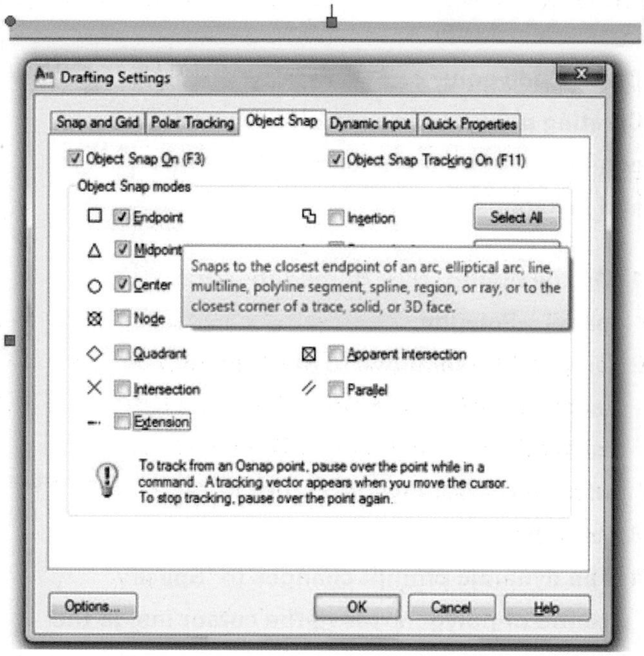

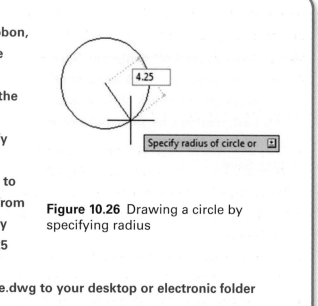

**Figure 10.25** The Drafting Settings dialog box

## Activity

**Draw a circle:**

1. With Drawing1.dwg open, on the ribbon, click Home tab – Draw panel – Circle dropdown menu – Centre, Radius.
2. Drag the crosshairs cursor down to the drawing area.
3. Click the left mouse button to specify the centre of the circle.
4. The dynamic input prompt asks you to 'Specify radius of circle.' Enter 4.25 from the keyboard and press the Enter key to create a circle with a radius of 4.25 (Figure 10.26).
5. Click Save and save drawing as circle.dwg to your desktop or electronic folder location.

**Figure 10.26** Drawing a circle by specifying radius

## Activity

Draw a polygon:

Creating polygons is a simple way to draw equilateral triangles, squares, pentagons and hexagons.

1. On the ribbon, click Home tab – Draw panel – Polygon.
2. Drag the cursor down to the drawing area. In the dynamic prompt, use the keyboard to enter 6 to specify the number of sides of the polygon (Figure 10.27).
3. Press Enter.
4. The dynamic prompt changes to 'Specify centre of polygon.' Move the cursor inside the circle to find its centre point. Click when you see the orange circle and the dynamic cursor 'centre', which indicates the centre of the circle (Figure 10.28).
5. The dynamic prompt changes to 'Enter an option' and you have a choice of two options. Click the 'Inscribed within circle' option (Figure 10.29).
6. The dynamic prompt changes to 'Specify radius of circle.' Enter 4.25 and press the Enter key to specify the radius of the inscribed circle. You have created a six-sided polygon inscribed within a circle (Figure 10.30).
7. Click Save.

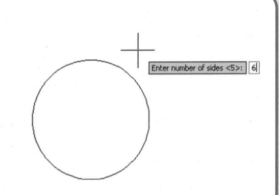

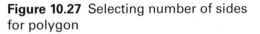

**Figure 10.27** Selecting number of sides for polygon

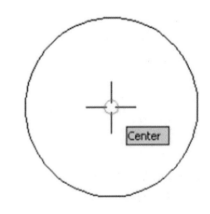

**Figure 10.28** Selecting centre of the circle

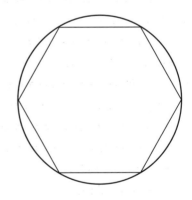

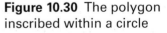

**Figure 10.30** The polygon inscribed within a circle

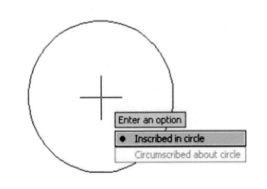

**Figure 10.29** Selecting inscribed within a circle

# Activity

Apply a hatch pattern: You can fill a closed boundary (any closed area such as a circle, rectangle or square) with a predefined hatch pattern; a simple line pattern using the current line type; or a custom hatch pattern. One type of pattern is called solid, which fills an area with a solid colour.

To use a predefined hatch pattern:

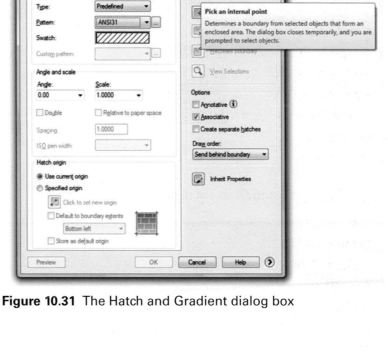

**Figure 10.31** The Hatch and Gradient dialog box

1. On the ribbon, click Home tab – Draw panel – Hatch.
2. In the Hatch and Gradient dialog box (Figure 10.31), click Add: Pick Points.
3. Click *inside* the polygon to select it. Do not select an edge of the polygon.
4. Press the Enter key.
5. In the Hatch and Gradient dialog box (Figure 10.33), Hatch tab, under Type and Pattern group box, specify the following using the dropdown menu:
   • Type: Predefined
   • Pattern: Steel
6. Click OK. The selected hatch pattern is applied to the polygon (Figure 10.34).

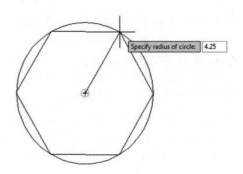

**Figure 10.32** Specifying the radius of circle

The scale can be reduced in the Hatch and Gradient dialog box if required.

7  Click Save, and then close the drawing.

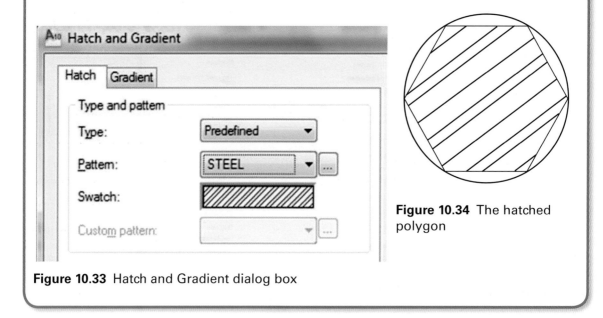

**Figure 10.34** The hatched polygon

**Figure 10.33** Hatch and Gradient dialog box

## Dimensions

An engineering drawing used in manufacturing should contain enough information so that the requirements for producing the engineered part or component are clearly defined. The engineering drawing needs to possess dimensions that specify features of size, position, location, surface texture and geometric control. Manufacturing parts or components to an exact size is not feasible. Therefore, tolerances are allowed on dimensions to indicate upper and lower limits of size. It is also important to ensure that the drawing is not overloaded with unnecessary dimensions. BS8888 covers the standards required for dimensions. Adhering to BS8888 will ensure a drawing is produced to professional standards.

There are several distinct parts to a dimension. These are the dimension text, dimension lines, arrowheads and extension lines. These are illustrated in Figure 10.35. The appearance and location of dimension properties can be altered using a dimension style manager.

A solid block with a circular hole is shown in Figure 10.36. To ascertain the exact shape of the object from a drawing we would need to establish the following dimensions:

- length;
- height;
- thickness;
- diameter of hole;
- position of the hole in relation to the sides of the block.

The axis of the hole is indicated by the intersection of two centre lines located at 30 mm from the left-hand side and 35 mm from the bottom of the block. These two surfaces are termed 'datum' or 'baseline' and the length and height have been measured separately from them. This avoids what is known as a 'cumulative error in measurement'. There is also no calculation necessary to determine the other dimensions.

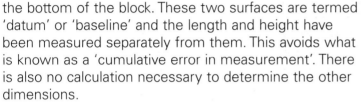

**Figure 10.35** Parts of dimensions

Dimensioning, therefore, is the process of adding measurement annotation to a drawing. You can create dimensions for a variety of object types in many orientations. The basic types of dimensioning are:

- linear;
- radial (radius, diameter and jogged);
- angular;
- ordinate;
- arc length.

Linear dimensions can be horizontal, vertical, aligned, rotated, baseline, or continued (chained). Some examples are shown in Figure 10.37.

The type of text used on dimensions should also be considered. No particular style is required, but characters should all be consistent on the same drawing. Capital letters are preferred to lower-case letters. The size of lettering is given as a minimum height, relating to drawing size, as shown in Table 10.4.

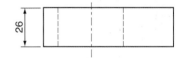

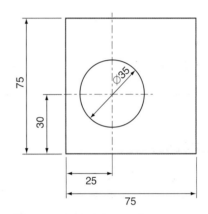

**Figure 10.36** Dimensions

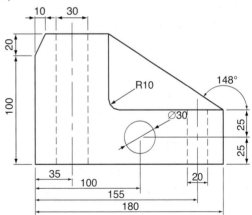

**Figure 10.37** Dimension styles

| Application | Drawing sheet size | Minimum character height (in mm) |
|---|---|---|
| Drawing numbers | A0, A1, A2 and A3 | 7 |
| Titles, etc. | A4 | 5 |
| Dimensions and notes | A0 | 3.5 |
| | A1, A2, A3 and A4 | 2.5 |

**Table 10.4** Drawing sheet sizes and sizes of lettering

## Activity

- Using a CAD package produce the drawing shown in Figure 10.38.
- Why is it important to avoid cumulative error in measurement?
- What is a datum?

## Viewing the CAD drawing

CAD software has a number of features to assist you when you want to view a specific region of a drawing in more detail – for example, if you were working on a large office building floorplan but you only wanted to view one room on screen. To do this, you use the Zoom command. This is similar to what happens when you zoom in and

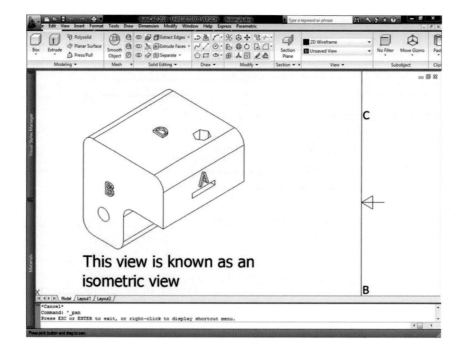

**Figure 10.38** Partially zoomed view of first-angle projection drawing

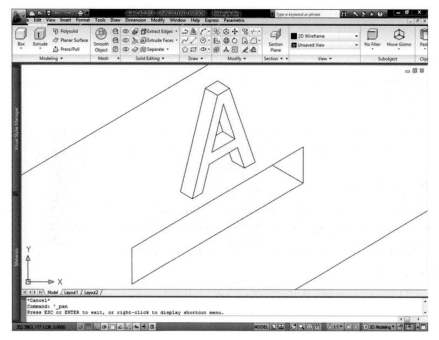

**Figure 10.39** Zoomed view on block detail

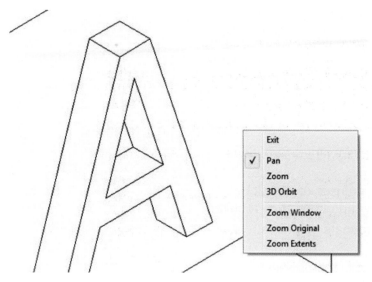

**Figure 10.40** The Pan and Zoom sub menu

out with a camera. The absolute size of the object you are looking does not change but the magnification of the view does. This is also useful when you are working on very small parts. You can zoom in and out to see an overall view of the work. Figure 10.38 shows a partially zoomed view of the first-angle projection drawing.

Figure 10.39 shows the same object with the magnification increased by using the Zoom command. There are a number of ways you can zoom in and out of a drawing using built-in software commands. You can draw a window around the area you want to zoom in to or zoom out to view all of the drawing with the Extents option. The easiest way is with an input device, such as a mouse with a scroll wheel. When you turn the scroll wheel, you may feel a small click. Each click or increment on the scroll wheel zooms by a factor of about ten per cent, but this can be changed in the settings.

Another feature of CAD software is the Pan command. This is similar to panning the view when looking through a camera. It does not change the magnification or location of the object; it just changes the view. To use the Pan command, you hold down the left mouse button and move the drawing around on screen. Both the Zoom and the Pan commands can be accessed through the dropdown menus or by direct input into the command line.

Once the command is active, you can right-click with the mouse to open a sub menu (Figure 10.40).

## Drawing checklist

It is easy to accidentally omit various items when creating engineering detail drawings. Before passing on your work, it is recommended that you work through the checklists below for each drawing.

### The general drawing

- Do projections conform to the relevant conventions, usually first or third angle?
- Have you used the minimum number of views necessary to accurately show the information required?
- Are the views laid out in appropriate positions relative to the size of paper?
- Has the title box been completed, particularly:
  - drawn by;
  - name of component;
  - date;
  - projection (first or third angle);
  - paper size;
  - scale.
- If required, has the material been specified?

## The geometry details

- Check to make sure that there are sufficient dimensions to manufacture the component.
- Check that positions and sizes of any features, such as holes, are clearly dimensioned.
- No dimension should appear more than once on the drawing; do any?
- Have the dimensions been laid out in consistent and clear positions, so that they are easy to read?
- Have all of the dimension lines been constructed with correct extension lines and gaps?
- Are the arrowheads all in the same style and the same size?
- Have dimensions relating to a particular feature, such as a hole, been grouped together on one view, if possible?
- Have appropriate line styles and line weights been used?
- Have any surface finish requirements been specified?
- Have any explicit tolerance requirements been specified?
- Have any required centre lines, break lines, etc. been used?
- Have any required general notes been added, such as additional general tolerances, finish specifications or specifications of special manufacturing processes?
- If sections have been used, do they conform to drawing conventions?

# Learning Outcome 3. Be able to modify engineering drawings using CAD commands

## Modifying engineering drawings using CAD commands

P5    P6

When we modify a CAD drawing, we make changes to it. This could be to rectify mistakes or to include new features on an existing drawing. To make changes to a manually produced paper drawing would require a great deal of time and effort. One advantage of CAD is that drawings can be modified easily on screen. There are a number of commands and functions available within CAD software for modifying drawings. Figure 10.41 illustrates the icons from the dropdown Modify menu. We will now cover some of the main concepts.

If you make a mistake while using CAD, you can backtrack and use the Undo command. This can be accessed from the pop-up menu by right-clicking with the mouse, from the dropdown menu or by entering the letter U in the command line. Undo has the effect of going back to the point just before you made the mistake.

There are many ways to delete objects drawn on screen. If you want to remove an object such as a line or circle then you can use the Erase command (Figure 10.42). A quick way of doing this is to select the object you want to erase, right-click the mouse and use the Erase option from the pop-up sub menu. You might actually erase the wrong object. If this happens, you can use a command called Oops to restore it.

A useful method of working with CAD is the use of multi-layers. If you can imagine multiple layers of transparent paper placed on top of each other, then you

**Figure 10.41** Modify icons

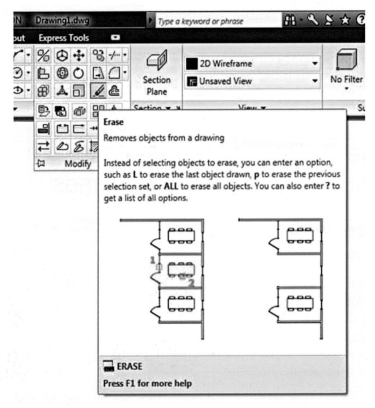

**Figure 10.42** Erase command

two lines
before fillet

two lines
filleted with
radius

two lines
filleted with
zero radius

original
objects

chamfer
distance zero

chamfer
distance
not zero

**Figure 10.43** Fillet command    **Figure 10.44** Chamfer command

have grasped the concept of layer control in CAD. Each of the layers underneath in the pile can be switched on or off, can be assigned different colours and line types, or can contain different parts of the same drawing. If we take this tracing-paper analogy further, we could draw the plans for a house. On layer 1 we could draw the plan view of the building plot, layer 2 we could use for a construction grid, layer 3 we could use for utilities (water, gas, electricity supplies) circuits, layer 4 could be the design for the ground-floor layout, layer 5 could be the location of the outside walls, and so on. This 'tracing' facility and the ability to modify, print or plot separate layers is an important drawing asset.

In addition to erasing objects on screen, you can modify them in other ways. It is possible to copy, move, rotate, stretch and scale objects. The Fillet command is used to turn square corners into rounds or arcs. It can be used on outside and inside corners. It does this by connecting the two objects to be filleted with an arc. You can specify the arc radius before you complete the Fillet command. If you specify the arc radius as zero, then the result is a fillet as shown in Figure 10.43.

The Chamfer command uses the same principle as the Fillet command but it applies a bevelled edge to the object. Figure 10.44 illustrates the action of the Chamfer command.

A good example of the efficiency of a CAD system when compared with a manually produced paper drawing is the use of the Array command. This enables multiple copies of an object to be placed on screen at a specified distance, location and orientation from the original object. To do this by hand on a paper drawing would take considerably longer. Figure 10.45 illustrates the Array command. The arrayed objects are shown in red.

You can shorten or extend objects to meet the

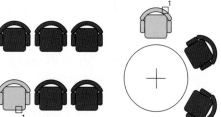

**Figure 10.45** The rectangular and polar Array command

edges of other objects. If you have objects overlapping, you can trim off the portion of the object that overlaps. Figure 10.46 illustrates how the Trim command has removed the overlapping sections of the diagonal line.

**Figure 10.46** Using the Trim command

# Learning Outcome 4. Be able to store and retrieve engineering drawings for printing/plotting

## Storing and retrieving engineering drawings for printing/plotting

P7   P8   M3

Since engineering drawings can be output to an electronic file or paper sheet, suitable storage, retrieval and duplication arrangements are necessary. Systems typically used are:

- filing drawings by hand into storage cabinets;
- microfilming or microfiche;
- computer-based storage.

Even for a relatively small drawing office, the management of paper drawings can require a great deal of time, effort and space. The drawings need to be filed in order with a suitable reference or indexing system to aid future retrieval if a modification or duplication were required. The preservation and security of the original drawing are also crucial.

Microfiche and microfilming used to be common methods of storing large volumes of data onto a photographic film format. With the advent of computer technology, however, these are rarely used now. Drawings stored using a paper-based or microform method can be scanned and converted to digital format for use on a computer.

As the use of CAD has become more widespread in engineering, then the requirement for storing large volumes of paper drawings has been reduced. Most drawings are now stored in an electronic file format on computer storage media. A drawing produced using CAD can be saved to the local computer hard disc, a network drive or portable storage devices. Data can be kept on removable media, such as CDs, DVDs and USB memory sticks. This offers many advantages over manually drafted drawings; however, the ease of duplication can present a security risk for commercially sensitive data.

With the majority of engineering drawings now being produced using CAD, an efficient and secure method of data storage and retrieval is required. A large drawing office may have to store hundreds or even thousands of engineering drawings in electronic format. To manage all of the drawings and keep track of any changes made to them, special software is used to store all the data in a centralised location. This allows the design team to easily share and manage information.

When saving a drawing produced using CAD, it is saved using the same basic procedure you would use with other applications such as Microsoft® Word. While you are working on a CAD drawing, it is good practice to save it at regular intervals. This will ensure you have a recent copy of your work in case the computer crashes or there is a power failure.

When you save an electronic file, you can assign any name you wish to it within reason (there is a limit of 256 characters). The part of the file name you cannot change is the file extension. This is the part after the full stop. The file extension for drawing files is .dwg; therefore a full file name for a CAD drawing could be drawing1.dwg.

Different software applications use different file extensions. Some examples are shown in Table 10.5.

File extensions for CAD drawings differ depending on the CAD application with which they were created. This can lead to compatibility issues when trying to work on a

| Application | File extension |
| --- | --- |
| Microsoft® Word | .doc |
| Microsoft® Excel | .xls |
| Windows Media® Video | .wmv |
| Shockwave Flash® | .swf |
| CATIA® CAD | .CATDrawing |
| Autodesk Inventor® 3DCAD | .ipt |

**Table 10.5** Examples of file extensions

**Figure 10.47** The Save As command

CAD drawing with a different CAD software package from the one with which it was created. There are, however, 'neutral' file formats, which allow the exchange of data between different CAD applications.

When saving a CAD drawing, a number of options are available. Figure 10.47 illustrates the File Save options for AutoCAD® 2010.

You will generally save drawing files with the .dwg file extension. However, your drawings can also be saved with the file extensions shown in Table 10.6.

| File type | File extension |
| --- | --- |
| CAD standards file | .dws |
| CAD template file | .dwt |
| AutoCAD® drawing file | .dwg |
| AutoCAD® ASCII file | .dxf |
| Back-up file | .bak |
| Automatically saved file | .sv$ |

**Table 10.6** File extensions for CAD drawings

The file extension you select depends on what you want to do with the file later. It is useful to create a back-up copy of each CAD drawing you produce. The CAD application can do this automatically in case of computer hardware problems. If a problem does occur with the original drawing, you can restore a copy from a back-up file. There is also a facility for the CAD application to automatically save the drawing file at regular intervals. This is useful if you forget to do the save yourself.

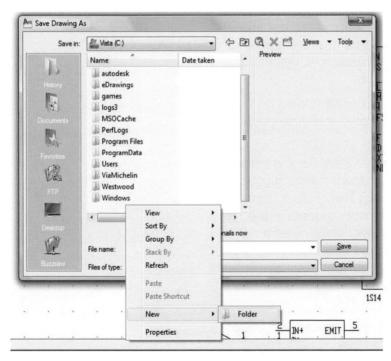

**Figure 10.48** Creating new folders 1

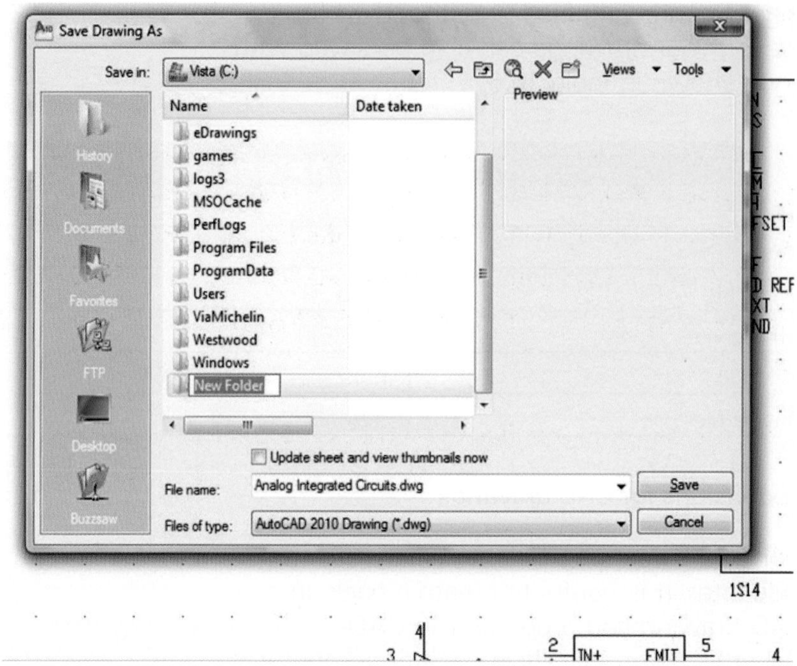

**Figure 10.49** Creating new folders 2

You may have many CAD drawing files to save, so it is good practice to organise them into specific directories or folders. You can create folders on your computer's hard disc or network drive and give them a name. You could name the folder by project, subject or customer name. For example, if you had produced a large number of diagrams for fluid-power, electrical and electronic circuits, you could organise them into separate folders.

In the following example, we will create folders on the computer hard disc (C drive) and save the drawings into the appropriate folders.

From the application menu, choose Save As and select the C drive in the dialog box dropdown menu. Right-click in the dialog box file list and select New Folder from the pop-up sub menu. A New Folder icon appears in the file list with the temporary name 'New Folder'. Type a name for the folder. Because the temporary name 'New Folder' is selected, the name that you type automatically replaces it. Either press Enter or click in the file list when you are finished (see Figures 10.48 and 10.49).

An alternative method is to click the Create New Folder button on the toolbar in the dialog box. A New Folder dialog box appears in the file list. Type a name for the folder. Either press Enter or click in the file list when you are finished.

## Activity

Organise a number of drawing files on a computer by creating folders and storing files in them. Suggestions for the drawing files and folder names are given below, but you can create your own or use drawings you have already produced. In this example, we would first create a folder on the C drive named 'Cadfiles' then create sub folders for each type of circuit.

These are the folders and drawing files to be created, organised and stored:

- Pneumatics:
  - Pneu_circuit_1.dwg
  - Pneu_circuit_2.dwg
  - Pneu_circuit_3.dwg
- Hydraulics:
  - Hyd_circuit_1.dwg
  - Hyd_circuit_2.dwg
  - Hyd_circuit_3.dwg
- Electrical:
  - Elec_circuit_1.dwg
  - Elec_circuit_2.dwg
  - Elec_circuit_3.dwg

You can follow this method to help you complete the activity.

1. Create the main folder:    C:\Cadfiles
2. Create the sub folders:    C:\Cadfiles\pneumatics
                              C:\Cadfiles\hydraulics
                              C:\Cadfiles\electrical
3. Save or move each drawing file into the appropriate folder.

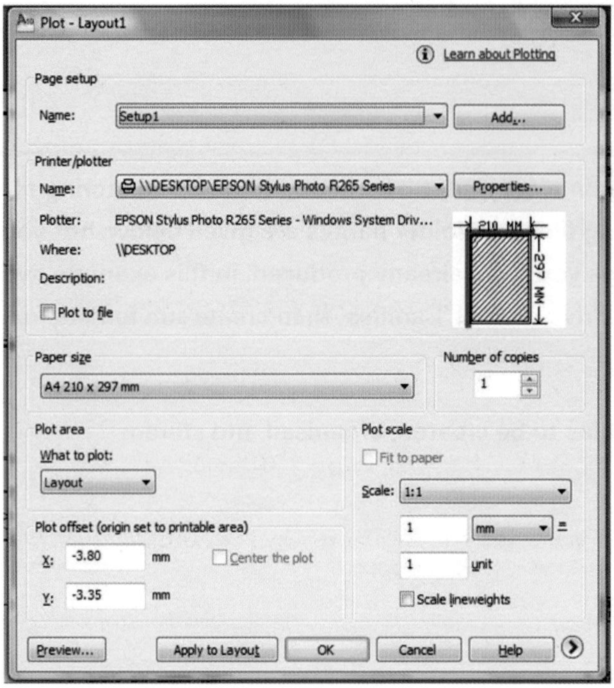

**Figure 10.50** A plot dialog box

To open a drawing for editing or printing, you can use the Open option from the quick access toolbar or application menu. You may want to do this to print out or edit an existing drawing. When CAD drawings are printed, they are generally referred to as being plotted out. Plotters are used to produce large-scale line drawings and traditionally they used a series of pens. More modern plotters are actually large-scale inkjet printers.

When you want to plot a CAD drawing, you use the plot dialog box (Figure 10.50). From here, you can assign the printer or plotter you want to use, select the paper size, orientation and number of copies required, apply a scale and decide what exactly you want to plot. You can plot out the limits of the drawing or just specific areas of it. A useful option is the provision of a print preview. This enables you to view what your intended plot will look like before you commit it to the plotting device or printer.

## Activity

Plot out a CAD drawing that you have produced. This could be a circuit diagram or a part design. You can use standard A4-sized paper for this exercise. If you have access to an A3 plotter or printer, then print out a drawing using this size. Practice using the options for 'What to plot' and plot out a selection of different views. For example, you could plot out the full drawing or just a specific section of it.

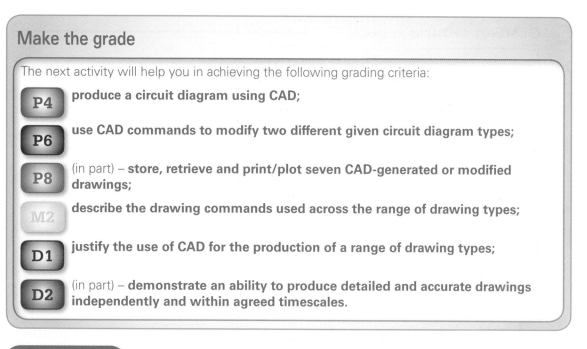

## Make the grade

The next activity will help you in achieving the following grading criteria:

**P4** produce a circuit diagram using CAD;

**P6** use CAD commands to modify two different given circuit diagram types;

**P8** (in part) – **store, retrieve and print/plot seven CAD-generated or modified drawings;**

**M2** describe the drawing commands used across the range of drawing types;

**D1** justify the use of CAD for the production of a range of drawing types;

**D2** (in part) – **demonstrate an ability to produce detailed and accurate drawings independently and within agreed timescales.**

## Activity

Produce a circuit diagram using CAD and modify two given circuit diagrams.

1. Produce a circuit diagram using CAD. For example, an electronic circuit or a pneumatic circuit.
2. Modify two circuit diagram types using CAD and save them to an electronic folder.
3. Print/plot out seven CAD-generated or modified drawings and include them with the report. You may have already completed this in the activity after Make the grade on page 236.
4. Describe the main drawing commands you used to produce and modify the circuit diagrams using CAD.
5. Justify why CAD is used to produce engineering drawings.
6. Complete all work in a reasonable time period and to agreed and appropriate standards.

This task is a practical activity but you will need to keep a record of the steps you used to complete it. You can use screen dumps by pressing the print screen (PrtScn) key and pasting the captured images into a Word document. Produce a report describing the methods used to complete the tasks detailed in this activity.

Obtain a witness statement or observation record from your tutor or supervisor to show you have completed this activity competently and achieved the criteria.

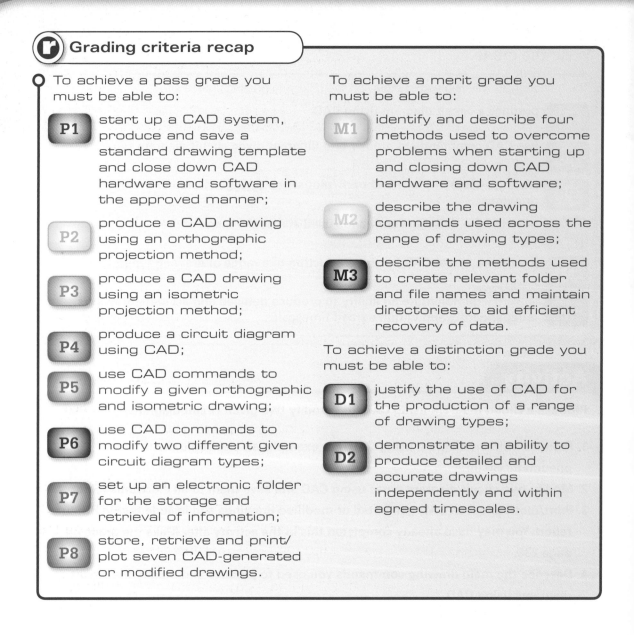

**Grading criteria recap**

To achieve a pass grade you must be able to:

**P1** start up a CAD system, produce and save a standard drawing template and close down CAD hardware and software in the approved manner;

**P2** produce a CAD drawing using an orthographic projection method;

**P3** produce a CAD drawing using an isometric projection method;

**P4** produce a circuit diagram using CAD;

**P5** use CAD commands to modify a given orthographic and isometric drawing;

**P6** use CAD commands to modify two different given circuit diagram types;

**P7** set up an electronic folder for the storage and retrieval of information;

**P8** store, retrieve and print/plot seven CAD-generated or modified drawings.

To achieve a merit grade you must be able to:

**M1** identify and describe four methods used to overcome problems when starting up and closing down CAD hardware and software;

**M2** describe the drawing commands used across the range of drawing types;

**M3** describe the methods used to create relevant folder and file names and maintain directories to aid efficient recovery of data.

To achieve a distinction grade you must be able to:

**D1** justify the use of CAD for the production of a range of drawing types;

**D2** demonstrate an ability to produce detailed and accurate drawings independently and within agreed timescales.

# Introduction to the unit

When manufacturers and machinists produce component parts, they need to decide which are the most suitable processes and machines to use.

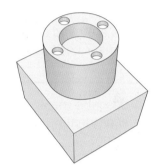

**Figure 14.1** Machined component

When components are manufactured, they are usually made using a number of different processes that follow one after another. In many cases, parts are shaped or forged by machines that produce the main shape; however, these shapes are not precisely accurate. Secondary machining is the process whereby a second machine is used to finish these parts so that they are accurate.

This unit examines the secondary machining techniques available and the tools needed to make these rough parts into accurate engineering components. This unit will help you to identify the range of machines available and to choose the most appropriate tools for manufacturing various components.

Four secondary machining techniques will be examined: turning, milling, drilling and grinding. The tools needed for machining components will also be discussed.

## Learning Outcomes

By the end of this unit you should:

- know how a range of secondary machining techniques is used;
- know how work-holding devices and tools are used;
- be able to use a secondary machining technique safely and accurately to make a workpiece;
- know about aspects of health and safety relative to secondary machining techniques.

# Grading criteria

| To achieve a pass grade you must be able to: | To achieve a merit grade you must be able to: | To achieve a distinction grade you must be able to: |
|---|---|---|
| **P1** describe how three different secondary machining techniques are used | **M1** explain why it is important to carry out checks for accuracy of features on components during and after manufacture | **D1** justify the choice of a secondary machining technique for a given workpiece |
| **P2** describe the appropriate use of three different work-holding devices for these different techniques | **M2** explain the importance of using the correct tooling and having machine parameters set correctly when machining a workpiece | **D2** compare and contrast three secondary machining techniques for accuracy and safety of operation |
| **P3** describe the appropriate use of three different tools for these different techniques | | |
| **P4** monitor and adjust the machining parameters to machine a given workpiece correctly and safely and to produce features as defined by the workpiece | | |
| **P5** machine a given workpiece safely and carry out necessary checks for accuracy | | |
| **P6** describe methods of reducing risk for the secondary machining technique used. | | |

# Learning Outcome 1. Know how a range of secondary machining techniques is used

## Secondary machining techniques: turning and milling

In this section, you will learn about two secondary machining techniques and how they are used:

- turning;
- milling.

### Turning

Turning is the process of removing material from a revolving workpiece by using a machine tool. The name of the machine that produces the part is a 'lathe', but there is a range of lathes that you will need to know about. These are:

- centre lathe;
- capstan lathe;
- turret lathe;
- automatic lathe.

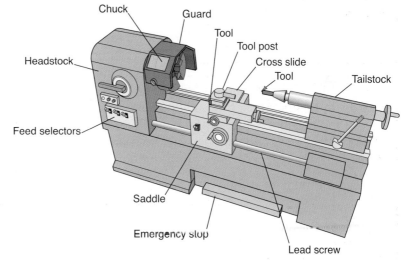

Chuck　　Guard
　　　　　　　　Tool
　　　　　　　　Tool post
Headstock　　　　Cross slide
　　　　　　　　　Tool　　Tailstock
Feed selectors
Saddle
Emergency stop
Lead screw

**Figure 14.2** Centre lathe

### Centre lathe

A centre lathe is the most widely used type of lathe. It produces cylindrical products and can be used to finish a material with screw threads, knurled finishes and profiles.

### Capstan lathe and turret lathe

A capstan lathe and a turret lathe are so similar that, in most cases, they are treated as the same thing. These are centre lathes with special parts added to them. These special parts help to make large numbers of components. A turret is a part of a machine that can rotate or revolve its tools, which saves time when changing tools.

## Automatic lathes

Some lathes are run automatically by computers. These can be used for very complicated components or for components that are mass-produced. Computer numerically controlled (CNC) machines have special systems that enable them to run automatically using computers.

# Milling

Milling is the process of using a machine to remove material to produce a product. The material is held in a vice or clamped to a table. The cutting tool is specially designed with cutting teeth that can cut though most engineering materials, including steel. The material is held in a work-holding device such as a machine vice. The material is then moved left and right or backwards and forwards. In this process, the cutting tool cuts out the shape of the product. There are many different methods for holding the work and various ways to machine material by milling. These will be examined throughout this unit.

The basic methods of milling are:

- horizontal milling:
  1. up-cut milling;
  2. down-cut milling;
- vertical milling.

## Horizontal milling

The term 'horizontal' refers to the relationship between the axis of the spinning cutter on the arbor and the ground. With a horizontal milling machine, the cutter spins on an axis parallel to the ground

The horizontal milling process is ideal for flat surfaces and cutting square edges. Different cutters can be used to produce long cuts with a constant profile.

*Up-cut milling:* with this method, the cutting tool's teeth rotate in a clockwise direction. The material moves towards the teeth, right to left. The speed at which the table moves is called the 'feed rate'. This is known as conventional cutting. It is easier on this machine as the tool pushes the material away, preventing backlash.

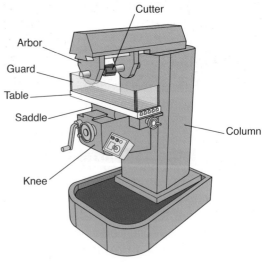

**Figure 14.3** Horizontal milling

*Backlash:* machine movement caused by the gaps between gears or other moving parts.

*Down-cut milling:* this method is called down-cutting because the cutting tool rotates anti-clockwise onto the workpiece. In order to use this method, a device known as a backlash eliminator is required. CNC machines do not have lead screws and so this method is used because it gives a better surface finish and is a more efficient method of cutting.

## Vertical milling

In the case of a vertical milling machine, the cutter spins in the vertical axis. (See Unit 3 Mathematics for Engineering Technicians.)

The workpieces and table move in exactly the same way as for horizontal milling, but the cutting tool is vertical, or perpendicular, to the table. It looks a little like a pedestal drill or bench drill but with a moveable table.

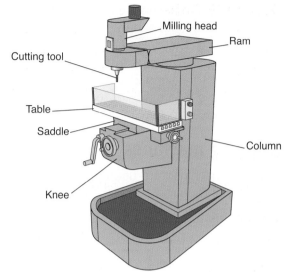

**Figure 14.4** Vertical milling

**Team Talk**

Aisha: **'What types of shapes can vertical milling machines produce?'**
Steve: **'A vertical milling machine can cut along edges or surfaces of material. It is great for making straight edges or a flat surface on the top of a workpiece. It can also produce slots, holes and keyways.'**

# Secondary machining techniques: drilling and grinding

In this section, you will learn about two secondary machining techniques and how they are used:

- drilling;
- grinding.

## Drilling machines

Drilling is the simplest type of secondary machining technique. Figure 14.5 shows a pedestal drill.

Drilling machines have a worktable that can move to locate the workpiece, but it does not move during drilling. The drilling machine is mainly used to drill holes. In an engineering workshop, holes from 1 mm to around 25 mm can be drilled. These machines are also used for reaming, which is an accurate way of producing a hole.

## Grinding machines

Grinding machines are used when extreme accuracy is required.

Grinders generally fall into two main types:

- surface grinders;
- cylindrical grinders:
  1. centreless grinding;
  2. profile grinding;
  3. thread grinding.

### Surface grinders

Surface grinders are used for very accurate machining. They use a spinning ceramic grinding wheel that rotates at a very high speed. A component that has been machined using a milling process can then be finished to a more accurate size using a surface grinder. Grinders have the advantage of being able to remove material that is very hard, which milling processes cannot do. They do not generally remove lots of material.

### Cylindrical grinders

A cylindrical grinder uses a high-speed, spinning ceramic grinding wheel to produce highly accurate components. The products produced are cylindrical. Where products have been machined using a lathe, a cylindrical grinding machine will finish these to a very accurate size. These machines can also finish boreholes to a high degree of accuracy.

*Centreless grinding:* this process is similar to cylindrical grinding but it does not use a spindle. It is used in mass production. The workpiece goes between the grinding wheel that cuts it and the regulating wheel that positions it.

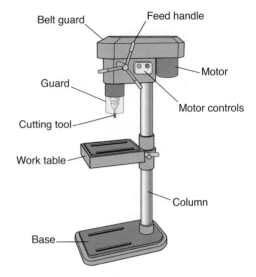

**Figure 14.5** Pedestal drill

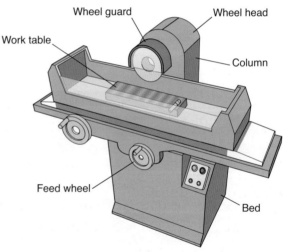

**Figure 14.6** Surface grinder

## Activity

Use the internet to find images of centreless grinding. Identify the regulating wheel, the workpiece and the grinding wheel.

*Profile grinding:* sometimes a more complex shape needs to be ground onto the surface of the workpiece. This requires cutting a profile into the grinding wheel, which will then cut this shape into the workpiece. This can be used for chamfers, radii, angles or more complicated profiles. A chamfer gives the edge of a hole a tapered entrance. It helps bolts or shafts entering the hole and it also stops the hole from having a sharp edge.

*Thread grinding:* a cylindrical grinder is used to produce threads. The grinding wheel is specially cut or 'dressed' using a diamond cutter. It can be used for internal or external threads.

### Make the grade

The next activity will help you in achieving the following grading criteria:

 **P1**  describe how three different secondary machining techniques are used.

**D1**  justify the choice of a secondary machining technique for a given workpiece.

## Activity

Explain how the following machines are used:

- centre lathe;
- vertical milling machine;
- surface grinder.

You should give a brief explanation of how the machine is operated, how the workpiece is held, and what the cutting tools are like. Also describe the types of shapes and cuts the machine can produce.

Use the example below (for a horizontal milling machine) to help you.

*'The horizontal milling machine has a spindle that turns, known as an arbor. Tools such as slab mills are mounted on this arbor. The tool spins but stays in a fixed position. The workpiece is usually held in a machine vice but can be clamped directly to the table. As the tool spins, the table is moved left to right so that the workpiece is moved under the spinning cutter. The table can be moved by hand using hand wheels or can be made to move automatically with a calculated feed to give a better surface finish. As the workpiece passes under the cutter, the cutter cuts away the material from the workpiece. The types of products that can be produced are long slots or vee-shaped cuts. If different tools are put together on the arbor, this will produce a few different profiles at the same time. This machine can cause a lot of swarf, so guards should always be in place and goggles should be worn.'*

# Learning Outcome 2. Know how work-holding devices and tools are used

## Work-holding devices for turning and milling

### Work-holding devices for turning

When products are being manufactured on lathes, special work-holding devices are used. These are needed to hold round bar, but lathes can also hold irregular-shaped components if those components are held correctly.

There is a wide range of holding devices for lathes, as follows:

- chucks;
- hard and soft jaws;
- collet chucks;
- drive plate and face plates;
- magnetic and pneumatic devices;
- fixed steadies and travelling steadies;
- four-jaw chucks.

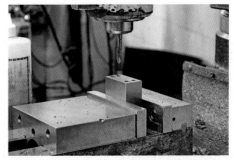

**Figure 14.7** Machine vice

### Chucks

A chuck is a special device that holds the material to be cut in place. It is similar to a vice, but all its jaws move. A three-jaw chuck, as the name suggests, has three of these jaws. A special tool known as a 'chuck key' is used to open and close the jaws.

**Figure 14.8** Chuck

This is a great method for holding round bar as the chuck automatically places the bar on the centre line of the centre lathe ready for machining.

### Hard and soft jaws

Chucks generally have hard jaws that cannot be machined. They do not produce cuts of extreme accuracy as the workpiece may be slightly off-centre. If soft jaws are used, the jaws can be machined so that they are perfectly on centre. These soft jaws are used where concentricity (see Team Talk below) is important.

## Collet chucks

A collet chuck is a special chuck used for producing accurate work. It will ensure that work placed into this chuck will be concentric to the centre of the centre lathe. A collet is a small holding device into which the workpiece fits; this collet then fits into the collet chuck and is tightened. Each collet suits only a small range of diameters, so each collet chuck is supplied with a range of collets.

66 Team Talk

Aisha: **'What does concentric mean?'**
Steve: **'If we drew two circles on top of each other with a compass, using the same centre point for the compass, the two circles would be concentric. We could say they had "concentricity".'**
Aisha: **'Do you mean like an archery target?'**
Steve: **'Yes. When we produce components on a centre lathe, this concentricity is extremely important.'**
99

## Drive plates and face plates

Sometimes an odd-shaped workpiece (for example motorcycle cylinder heads or a brake hub) needs to be machined on a centre lathe but a chuck is not appropriate. A face plate is a flat round plate with slots, so odd-shaped workpieces can be clamped to it. This can have the disadvantage of being out of balance, which means that the centre lathe would vibrate as the spindle speed increased.

## Magnetic and pneumatic devices

To save time clamping, if a material is magnetic, it can be placed on a face plate that is magnetised. A lever changes the magnetic chuck from non-magnetic to magnetic, which then holds the workpiece. These magnetic chucks are often used on cylindrical grinders. One disadvantage of these is that they cannot take high cutting forces.

Pneumatic devices can clamp workpieces using air pressure. This is quick and easy.

## Fixed steadies and travelling steadies

When a bar protrudes from a chuck, it can be quite strong and will not bend. However, when a bar is long or thin, it can deflect as it is machined. A special device called a fixed or travelling steady can be used to support a long workpiece; this has three jaws that gently rest on the workpiece. This device is strong enough to prevent the workpiece from moving during machining.

## Four-jaw chucks

A four-jaw chuck looks similar to a three-jaw chuck, but as its name suggests it has four jaws. These jaws do not move together and are used to hold workpieces where the part to be machined is off-centre to the centre lathe. This can result in the workpiece being 'out of balance', which may cause the centre lathe to vibrate at high speed.

# Work-holding devices for milling

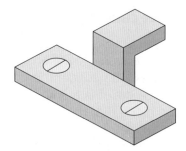

**Figure 14.9** Fixed steady

When products are manufactured using milling machines, a range of clamping methods can be used. These are as follows:

- direct clamping;
- machine vices;
- chucks;
- angle plates and vee blocks;
- fixtures;
- indexing heads;
- rotary tables;
- pneumatic and magnetic clamps.

**Figure 14.10** Part being milled

## Direct clamping

The simplest way to hold a workpiece on a milling machine is to clamp the work directly onto the milling machine table. Special slots are machined into the table to allow clamps to slide into position where they are tightened. Small steps or adjustable blocks sit under the clamp to keep the clamps level.

## Machine vices

This is the most commonly used method of holding on a milling machine. Machine vices hold the workpieces in a similar way to a bench vice but are extremely accurate and robust. They can hold square and rectangular workpieces easily. They can be set 'square' using a dial test indicator; this gives very accurate positioning.

## Chucks

A chuck similar to that used on a centre lathe can be clamped directly to the milling machine table. This allows round bar to be held vertically.

## Angle plates and vee blocks

Sometimes a workpiece needs to be set vertically or perpendicular to the worktable. In this case, an angle plate is used. As with the table, there are slots in the angle plate that allow the workpiece to be clamped. Where round bar is to be held, a vee block can be used. This is a more secure method of holding round bar than a chuck. Look at angle plates and vee blocks in detail in Unit 18.

## Fixtures

A fixture is designed for mass production. If the workpiece is an odd shape that is not easily held by a vice or chuck, a fixture can be made which is specially designed to locate the workpiece. The clamps are built into the fixture, in set positions, so they are quickly and easily clamped

## Indexing heads

An indexing head is a device specially designed to rotate bar through a pre-determined angle. The round bar is held horizontally and machined. The indexing head is then rotated using a handle and a series of slots at exact locations revolve the bar through the angle needed. The bar is then machined in the new location. A hexagon or octagon could be produced using this method.

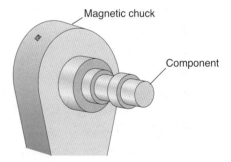

Figure 14.11 Indexing head

## Rotary tables

A rotary table is used in a similar way to an indexing head but it has divisions around the edge to indicate the angle that has been moved. On CNC machines, these tables can move automatically.

## Pneumatic and magnetic clamps

To increase work capacity, automatic clamps such as pneumatic or magnetic clamps can be used. These will save an operator time when tightening and un-tightening bolts.

# Work-holding devices for drilling and grinding

## Work-holding devices for drilling

Drilling uses similar processes to milling and can use most of the clamping systems used for milling machines, as described in the preceding section. These are:

- direct clamping;
- machine vices;
- chucks;
- angle plates and vee blocks;
- fixtures;
- indexing heads;
- rotary tables;
- pneumatic and magnetic clamps.

These systems are detailed in the preceding section.

In general workshop use, handheld vices, vee blocks and direct clamping are most often used.

### Handheld vices

A very commonly used device is the handheld vice. It is useful where small holes are to be drilled and where speed of manufacture is important. They are not as accurate as machine vices. These vices have a flat base, which allows them to be clamped to the worktable. When students are learning to use these vices, it is always best practice to clamp them to the worktable.

## Work-holding devices for grinding

As we have seen in an earlier section of this unit, there are two basic grinding machines: surface grinders and cylindrical grinders. We will look at the two different machines separately.

### Work-holding devices for surface grinders

The main method of work holding on surface grinders is to use a magnetic table (also known as a magnetic vice).

The workpiece is placed onto the magnetic vice. The vice is then magnetised by the use of a lever. This will only work if the workpiece is a magnetic material such as steel. The workpiece is surrounded by accurate blocks of steel to prevent the wheel pushing it over. This is known as blocking. For non-magnetic materials, a vice can be held onto the magnetic table. Vee blocks and angle plates can also be used; these are clamped by magnetism to the magnetic table.

## 66 Team Talk

Aisha: **'How does the grinder table know how far to go backwards and forwards?'**
Steve: **'The distance that the table moves left to right is set by the use of control stops. These are small levers that are adjusted to make the table go from left to right, then right to left, and so on.'** 99

Figure 14.12 Cylindrical grinder

### Work-holding devices for cylindrical grinders

A cylindrical grinder operates in a similar way to a centre lathe: the workpiece revolves around a centre. Therefore, chucks and face plates can be used in the same way as discussed already.

A very common method of holding on a cylindrical grinder is to use a magnetic chuck. These are flat discs that are used like chucks but they do not have any jaws. A chuck key is turned, causing the flat surface to be magnetised. This can only be used when the workpiece has a very flat, wide diameter surface to slide on the surface of the vice. Of course, the workpiece needs to be a magnetic material such as steel.

### Make the grade

This activity will help you in achieving the following grading criterion:

**P2**  describe the appropriate use of three different work-holding devices for these different techniques.

## Activity

Explain how the following work-holding devices are used:

• **three-jaw chuck (used on a centre lathe);**
• **machine vice (used on a vertical milling machine);**
• **magnetic table (used on a surface grinder).**

You should give a brief explanation of how each work-holding device is used and the types of shapes or materials that can be held in these devices.

Use the example below for a four-jaw chuck (used on a centre lathe) to help you.

*'A four-jaw chuck is used on a centre lathe. There are four separate jaws that move independently. This means that a four-jaw chuck can hold square bar and octagonal bar easily. It can also hold round bar, but because the jaws move separately, the round bar can be set on the centre line or eccentrically. When using round bar, the workpiece does not automatically sit on the centre line of the lathe. This means it is not used for speed. It is used to set workpieces accurately. Where accurately finished workpieces are used, this type of chuck may damage the surface. A collet chuck would be a better alternative.'*

# Tools for turning and milling

P3

## Turning tools

Centre lathes use a number of specialist tools, as well as tools that are used in other processes. We will look at these in two groups:

- turning tools – specifically used on lathes;
- drilling, reaming and tapping – used in other processes.

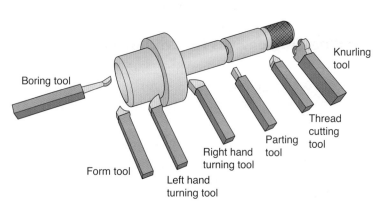

**Figure 14.13** Turning tools

Turning tools come in a number of basic types and are as follows:

- left-hand cutting tool;
- right-hand cutting tool;
- facing tool;
- form tool;
- parting-off tool;
- thread-cutting tool;
- boring tool;
- knurling tool.

### Left-hand and right-hand cutting tools

These are used to cut along the length of a bar, which creates a diameter.

### Facing tool

This is a tool used to cut along the front surface of the bar to give a flat surface.

### Form tool

This is a tool that can be ground to a shape, such as a radius or angle, or in fact any odd shape. This tool then creates the same shape on the outside of a workpiece.

### Parting-off tool

This is a thin, rectangular-shaped tool used to cut through the full diameter of the bar until the component being manufactured drops off the end of the bar.

### Thread-cutting tool

This tool has a very accurate vee-shaped cutting edge, which forms the vee shape of a thread.

### Boring tool

This is a special tool used to cut diameters on the inside of components.

### Knurling tool

This is a special tool that forms a rough surface in the form of small diamond shapes. This type of surface finish is used where grip is needed. It could be used where handles are produced for mallets, screwdrivers and die holders.

Each of these tools is held in the tool post of the lathe. It is important that they are set on the centre height of the lathe.

## Milling cutters

Milling cutters come in a range of types, but vertical and horizontal milling machines use special cutters for the specific type of machine. Therefore, the cutter types are organised into two groups: vertical and horizontal milling cutters.

## Vertical milling cutters

Vertical milling cutters include:

- slot drills;
- end mills;
- face mills.

*Slot drills:* these are basic milling cutters with two or sometimes three teeth. These can be used to cut profiles but are especially used to go straight into metals to form flat-bottomed holes, slots or pockets.

*End mills:* these cutters have four or more teeth. They cannot go straight into metal but are used mainly to cut profiles. The more teeth they have, the faster they will cut around the profile.

*Face mills:* the main function of these tools is to cut the top surface of a workpiece. They are very efficient and can cut a large area because they can be big in diameter

**Figure 14.14** Endmill

**Figure 14.15** Side face and cutters

## Horizontal milling cutters

There are two main horizontal cutters, as follows:

- side and face cutters;
- slab mills.

*Side and face cutters:* usually of a large diameter and thin in width, these cutters can cut edges of workpieces and are excellent for long slots in workpieces

*Slab mills:* these are heavy cylinders with many teeth, designed to remove lots of metal from the top surface of the workpiece.

# Tool materials

Turning tools and milling tools are available in a range of materials and forms, which are used in both processes. These include:

- high-speed steel (HSS);
- cemented carbides and sintered carbides;
- indexible tips.

## High-speed steel (HSS)

This is a common material that is relatively cheap. It contains 0.65 to 0.8 per cent carbon and 3.5 to 4.0 per cent chromium, plus additions of tungsten and other metals. This cutting material is suitable for cutting mild steel, but over heating will make the cutter soft. (For more information see Unit 8 Selecting engineering materials.)

### Cemented carbides and sintered carbides

Made from tungsten carbide or a mixture of tungsten carbide and other materials, this is an expensive material and not very strong. Small 'tips' are made from this material and screwed to a steel shank, which gives it strength.

### Indexible tips

Indexible tips are excellent for CNC machines as the tip can be thrown away and the replacement will be in exactly the same position as the previous tip. In some cases, the tips may be square, triangle or diamond-shaped, so they can revolve, giving separate cutting edges. These are known as indexible tips.

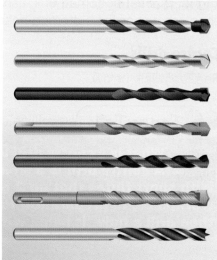

**Figure 14.16** Twist drills

# Tools for drilling and grinding

### Drilling tools

Although drilling machines are very simple, they can produce a range of features because they use a wide range of tools. These include:

- twist drills and flat-bottomed drills;
- centre drills;
- counterboring and spot facing tools;
- countersinking tools;
- machine reamers;
- threading taps.

### Twist drills (or drill bits) and flat-bottomed drills

These are the most recognisable and common of all cutting tools. The long, pointed cutter with two teeth or 'flutes' is used to drill holes. These are made from high-speed steels. In a workshop, they range from 1 mm diameter to around 50 mm, but they can be much bigger in diameter using specialist machines. They produce round holes all the way through a workpiece or only part way through (known as a 'blind hole'). The bottom of the hole will have a cone shape. A twist drill with a flat end,

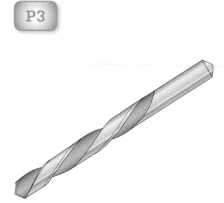

**Figure 14.17** Drill with two teeth or flutes

known as a flat-bottomed drill, can be used to create counterbores or flat-bottomed holes.

### Centre drills

See Team Talk below.

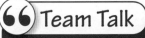

**66 Team Talk**

Aisha: **'What is a centre drill used for, Steve?'**
Steve: **'A centre drill is a small, solid drill bit, with short flutes that go 2 or 3 mm into the workpiece, in preparation for a twist drill to then drill a hole. It helps with accuracy.'**

99

### Counterboring and spot facing tools

A counterboring tool is designed to drill a larger hole into an existing hole, known as a counterbore. This is used to allow nuts or bolt heads to sit below the surface of the workpiece. A spot facing tool will do a similar job but generally only goes 1 or 2 mm into the surface of the workpiece to create a flat surface.

### Countersinking tools

These are special tools that create chamfers around the top edge of a hole. This makes edges safe from sharp corners and prevents the edges of holes becoming damaged.

### Machine reamers

A machine reamer can be described as a very accurate twist drill. It is used after a twist drill has produced a hole and is slightly bigger than the twist drill. The diameter of the hole produced is extremely accurate and smooth.

### Threading Taps

Taps produce internal threads. They are basically screws threaded with slots up their length to allow the cutting. They come in a wide range of sizes and thread types. The taps can be held in a drill chuck, which is turned by hand to produce the thread into a pre-drilled hole. This ensures thread alignment.

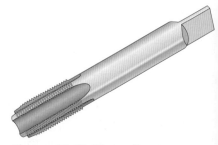

**Figure 14.18** Threading tap

# Grinding tools

Grinder cutting tools are referred to as 'grinding wheels' and have two important elements: what they are made from, and their shape.

Grinding wheels are complex but, put simply, they have a hard ceramic that forms the gritty texture. These materials can be:

- aluminium oxide;
- silicon carbide;
- diamond;
- cubic boron nitride.

This material is held together with a grinding wheel bond. The types of bond include:

- vitrified;
- resinoid;
- silicate;
- shellac;
- rubber;
- oxy-chloride.

## Hard and soft wheels

Grinding wheels can be 'hard' or 'soft'. These terms relate to the bond type and not to the hardness or softness of the ceramic.

## Grinding wheel shapes

All grinding tools are made from ceramic rather than metals used in other processes. The cutting tools vary in shape and are as follows:

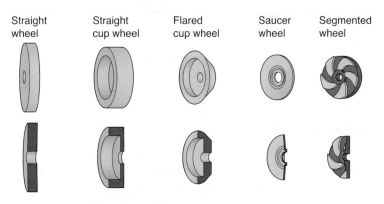

Figure 14.19 Grinding wheel shapes

- straight wheel;
- straight cup wheel;
- flared cup wheel;
- saucer wheel;
- segmented wheel.

*Straight wheel:* this is the most common style of wheel on surface, centreless and cylindrical grinders. In workshops, they range from around 150 mm diameter to

300 mm diameter. They reduce in diameter each time they are cleaned and dressed flat. They range from around 10 mm to 40 mm wide.

*Straight cup wheel:* the cup wheel is designed to cut on the side face of the wheel. This will cut the vertical sides of a workpiece.

*Flared cup wheel:* similar in use to a straight cup wheel but with tapered sides. This allows for a stronger wheel at bigger diameters.

*Saucer wheel:* this wheel looks like a saucer and is used to grind sharp corners on high-speed steel-milling cutters.

*Segmented wheel:* these are wheels with gaps, which can look like cutting teeth. They are used at very high speed, usually to cut marble, stone or other ceramics.

## Make the grade

The next activity will help you in achieving the following grading criterion:

**P3** describe the appropriate use of three different tools for these different techniques.

## Activity

**Explain how the following machines are used:**

- **left-hand cutting tool (used on a centre lathe);**
- **straight wheel (used on a surface grinder);**
- **slot drill (used on a vertical milling machine).**

**You should give a brief explanation of how each tool is used and what features it can produce.**

**Use the example below for an end mill (used on a vertical milling machine) to help you.**

*'An end mill has three or more teeth and is used in vertical milling. It can cut the sides of workpieces and, on CNC machines, it can cut profiles. The more teeth that an end mill has, the faster it can cut out a profile. When a cutter has four teeth, such as an end mill, it cannot go vertically down into a workpiece. In this case, a slot drill would be used. The end mill can only go 1 mm into a workpiece, so can be used to spot face a surface or counterbore a hole where the first hole has already been drilled.'*

# Learning Outcome 3. Be able to use a secondary machining technique safely and accurately to make a workpiece

## Turning and milling machining parameters

In this section, we will look at the parameters of machining. This means how fast tools or workpieces need to spin or move in order to machine workpieces effectively. For each machine, we will look at the following parameters:

- position tools to workpiece;
- cutting-fluid flow rate;
- machine guards;
- threading/profiling/tapering;
- feeds;
- speeds;
- depth of cut for roughing and finishing.

Figure 14.20 Turning on a lathe

### Turning

In this section we will use mathematics to determine values. You may refer to Unit 3 Mathematics.

### Position tools to workpiece

The workpiece in a centre lathe may be held in a number of ways using different devices. The most common method is a three-jaw chuck.

The three-jaw chuck holds round bar, which spins about its centre. There is also a range of turning tools but, in general, the turning tool is held in the tool post. The tool is positioned to the right of the workpiece between the operator and the workpiece as it spins. Removing a small amount of material from the outer surface creates a new diameter. When this is measured, the dial on the cross slide is set to zero and gives a position of the tool.

### Cutting-fluid flow rate

Coolant is a white, oil-based liquid used to cool the cutting tool. It can also help flush out swarf or debris from workpieces. The coolant on centre lathes is a long, thin pipe that is moved by hand to aim at the cutting tool. When drilling on lathes, the coolant pipe should be aimed directly into the hole being drilled in order to keep the drill tip cool.

## Machine guards

Machine guards are essential on lathes. Lathes should not be operated unless guards cover all moving areas. On centre lathes, the chuck is guarded as it spins. Often guards have a special micro switch that will not allow a machine to start without the guard being in place.

## Threading/profiling/tapering

Lathes have the ability to create screw threads. To create a screw thread, as the chuck spins it is connected to a special screw known as a lead screw by a series of gears. This results in the tool moving left to right relative to the speed of the chuck. This forms a thread when a pointed tool is mounted in the tool post.

For different threads, different gears are used. These gears are selected by a series of coded levers that relate to different threads.

## Feeds

The 'feed' is how fast the cutting tool moves left to right, right to left or across the face of the workpiece. It is measured in mm per revolution (of the chuck). If in one revolution of the chuck the tool moves 2 mm, this would be 2 mm/rev. Typically, smooth finishes may be 0.01 mm/rev. The bigger this number, the coarser the finish.

## Speeds

Cutting speeds relate to how fast a cutter can cut a material. The cutting speed for steel is around 20–30 metres per min (m/min); aluminium is around 100 m/min. However, this is for cutting along a straight line as the chuck is spinning. This can be converted into how fast the chuck spins in revolutions per minute (r.p.m.). The following formula is used to determine r.p.m., or 'N' (number of revs). D is the diameter of the material to be cut.

Formula for revs per minute of chuck (N):

$$N = \frac{\text{cutting speed} \times 1000}{\pi \times D}$$

## Depth of cut for roughing and finishing

In order to remove material, we need to wind in the cross slide an amount to take a cut. This is known as the 'depth of cut'. At this stage, you do not need to calculate the depth of cut for each material, but you do need to know

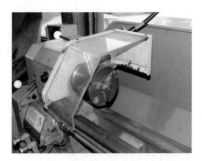

**Figure 14.21** Machine guard

that in general the smaller the cut, the smoother the finish. For speed of machining, a deep cut is needed, but the material must not bend under the force of the cut. See Figure 14.2 on page 264 which shows a cross slide

# Milling

## Position tools to workpiece

We will use vertical milling to demonstrate this. The centre of a milling cutter is used to position the workpiece with the cutting tool. An approximate method is to touch the workpiece with the cutter as it spins. The dial that relates to the machine table is set to zero. The tool is then moved vertically upwards. If we then move half the diameter of the cutter towards the workpiece, this locates the centre of the cutter to the edge of the workpiece.

A special device called an 'edge finder' is used to touch on the edge of the workpiece without marking it. For CNC machines, a special electronic 'probe' is used to give a digital position of the workpiece.

## Cutting fluid flow rate

Coolant is used in the same way as described for turning.

## Machine guards

All moving parts must be guarded. This can make it awkward to see what is being milled. Modern milling machine guards use strong, transparent plastic sheets fixed to metal hinges. These guards give good visibility, have good overall strength and are fitted with micro-switch safety mechanisms.

## Feeds

Milling cutters have one or more cutting edges (teeth). The more teeth, the greater the table feed rate.

Table feed in mm/min = r.p.m. of the cutter $\times$ number of teeth of cutter. $\times$ feed per tooth

## Speeds

The 'speed' refers to the speed of rotation of the spindle. This calculation is the same as that used in turning, where N is the number of revs and D is the diameter of the material to be cut:

$$N = \frac{\text{cutting speed} \times 1000}{\pi \times D}$$

**Figure 14.22** Milling

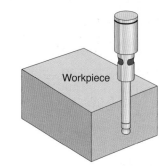

**Figure 14.23** Edge finder

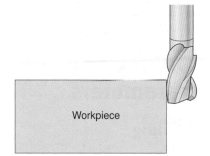

**Figure 14.24** Finding an edge with a tool cutter

**Figure 14.25** CNC probe

> ### Worked Example
>
> A high-speed steel cutter with four teeth is spinning at 400 r.p.m. It cuts low-carbon steel with a feed per tooth of 0.08 mm. Calculate the feed of the table.
>
> Solution:
>
> Table feed in mm/min = r.p.m. of the cutter × number of teeth of cutter × feed per tooth
> Table feed in mm/min = 400 × 4 × 0.08 = 128 mm/min

## Depth of cut for roughing and finishing

Standard milling cutters are generally about twice as long as their diameter. For instance, a 20 mm diameter cutter is about 40 mm long. Most cutters can cut full depth, but this can very easily snap the cutter. When machining profiles, the full depth of cut is often used. This is because only part of the cutter's diameter is used.

The more stable and robust the machine is, the better the cutter will remove metal. CNC machines are designed to withstand heavy cuts.

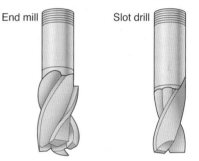

**Figure 14.26** Standard milling cuts

# Drilling and grinding machining parameters

P4

## Drilling

### Position tools to workpiece

In general, drilling machines locate the hole they are drilling by a centre punch indent created in the position to be drilled. There is no setting of table positions.

### Cutting-fluid flow rate

Most drilling machines are not fitted with a coolant pipe. For most occasions, each hole will be cooled by the air around the hole or a small bottle of coolant can be squirted manually into the hole as it is drilled. This method should only be performed by a fully trained operator. For plastics and woods, coolant is not used.

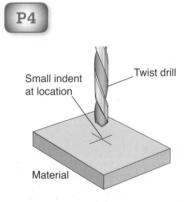

**Figure 14.27**

### Machine guards

All drilling machines are fitted with machine guards. These must be used at all times.

## Feeds and speeds

Basic drilling speeds can be found in Table 14.1.

| Drill diameter | Mild steel | High-carbon steel | Brass | Aluminium |
|---|---|---|---|---|
| 0–3 mm | 2,500 | 2,000 | 4,800 | 5,000 |
| 4–8 mm | 1,200 | 1,000 | 2,400 | 2,500 |
| 8–10 mm | 950 | 600 | 2,000 | 2,000 |

**Table 14.1** Basic drilling speeds (in revolutions per minute (RPM))

The feed rates for drilling are dependent upon the diameter of the twist drill. Basic milling feeds are shown in Table 14.2.

| Drill diameter | Feed rate (mm per minute) |
|---|---|
| 0–3 mm | 0.05 |
| 4–8 mm | 0.08 |
| 8–10 mm | 0.15 |

**Table 14.2** Basic milling feeds

## Depth of cut for roughing and finishing

Drilling manually is often based upon the operator's feel for how the drill is cutting. As a guide, the drill should go about the same depth as the diameter of the cutter (for a 10 mm diameter twist drill it should go 10 mm deep), then it should be removed from the drilled hole. A further depth should then be drilled, which should be about the same as the diameter of the hole (now it should go 20 mm deep, and so on). On CNC machines, this can be programmed into the machine.

> ## 66 Team Talk
>
> Aisha: **'Why do we need to keep going in and out with the drill?'**
> Steve: **'This will give the drill time to cool. Also, the swarf will be ejected and coolant can enter the drilled hole to cool the workpiece.'**
> 99

# Grinding

## Position tools to workpiece

Grinding is more accurate than turning or milling, and more dangerous. For your work, it must only be used under supervision. Setting the wheel onto the workpiece takes a lot of skill, as small movements will make a big difference to the cut of the wheel.

On surface grinders, the workpiece is held in place as described previously. The grinding wheel is slowly lowered onto the workpiece as it moves left to right and so on.

As the wheel touches the workpiece, small sparks will be seen. This is the position of the top of the workpiece. The wheel is lowered one hundredth of a millimetre at a time as it goes backwards and forwards over the workpiece. When the full surface has been ground, the dial showing the height of the wheel is set to zero.

**Figure 14.28** Coolant

### Cutting-fluid flow rate

The coolant must be on at all times during the grinding process. Whenever the wheel is stopped, the coolant must not run onto the wheel as this will 'load' the wheel and may put the wheel out of balance.

### Machine guards

The grinding wheel has it own guard that encloses the wheel. Guards at the front of the moving table protect the operator from grinding dust.

**Activity**

**Produce the component shown in Figure 14.29 using a vertical milling machine.**

- **Use appropriate work-holding devices and tooling.**
- **Show calculation for speeds and feeds.**
- **Use coolant.**

# Features of the workpiece for turning and milling

## Features created by the turning process

The turning process can produce the following features:

- basic turning features;
- holes;

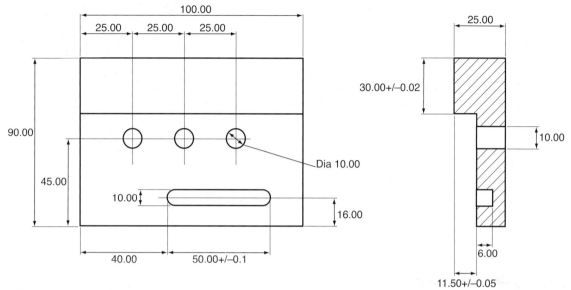

*All measurements in millimetres

**Figure 14.29**

- threads;
- eccentric features;
- knurling;
- profiles;
- parting off and grooves.

## Basic turning features

A centre lathe produces cylindrical shapes. At its most basic, it will produce cylinders. It can also cut across the surface of a cylinder to form a flat surface. It can cut different diameters on the same product.

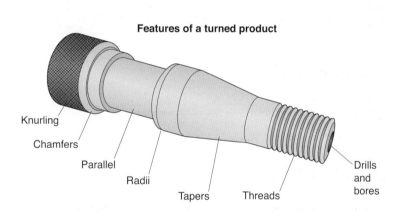

**Features of a turned product**

**Figure 14.30** Features created by the turning process

## Holes

The centre lathe can produce drilled, reamed or tapped holes but only easily if the hole is on the centre line of the workpiece. The hole will be parallel to the outside diameter of the workpiece. Holes off-centre can be drilled, but this is difficult to set up and usually a milling machine is a better production method.

## Threads

All types of threads can be cut on a centre lathe – those with standard measurement such as metric threads but also individually designed threads. It can cut vee-shaped threads, square threads or acme threads.

## Eccentric features

Where diameters are eccentric (the diameters are not on-centre), a four-jaw chuck is usually used.

## Knurling

This is a special feature that requires a special tool. A knurling tool, which has two wheels with sharp grooves, is pressed against the workpiece as it revolves. This forms a knurl pattern.

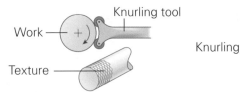

**Figure 14.31** Knurling

> ## 66 Team Talk
>
> Aisha: **'What does a knurl do, Steve?'**
> Steve: **'A knurl gives a product a surface finish of minute diamond shapes, which will give the product grip when it is used.'** 99

## Profiles

On centre lathes, form tools, such as chamfering tools and radius tools, can create special features. Complicated profiles are difficult to produce accurately on conventional centre lathes. CNC lathes can be programmed to follow very complicated profiles.

## Parting off and grooves

A parting-off tool is a long, thin-bladed tool used to cut deep into the workpiece until it cuts all the way through; the workpiece then falls away from the main material. Grooving tools are similar but are used to cut only part way into the workpiece, which creates a deep cut known as a groove. Grooving tools can move left to right and therefore can produce wide grooves. Often a small radius is formed on the corners of a grooving tool, which in turn creates a radius in the bottom of the grooves.

# Features created by the milling process

The milling process can produce the following features:

- flat, square and parallel faces;
- steps and shoulders;

- angular faces;
- slots;
- holes;
- profiles;
- indexed patterns.

## Flat, square and parallel faces

The basic features created by milling machines are flat surfaces and edges 90 degrees to these flat surfaces.

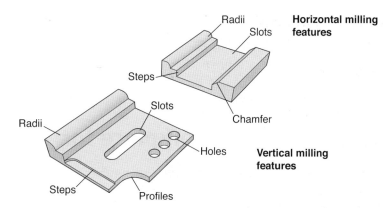

**Figure 14.32** Features created by the milling process

## Steps and shoulders

If one area needs to be lower than another, this is known as a step or shoulder. A workpiece can have as many steps or different levels as a drawing requires.

## Angular faces

Most milling machines have a milling head, which can be moved through different angles. This allows cutters to be tilted at an angle to create an angled feature on a workpiece. In some cases, the table can move through angles to create angled features.

## Slots

Milling machines can drill straight into a workpiece, similarly to a drilling machine. They can also move horizontally once the tool has reached its depth, using a special cutter known as a slot drill. Slots almost as long as the machine table can be produced. These can be open-ended or closed slots, such as a keyway. Tee-shaped slots can also be produced; these are used to hold clamps.

## Holes

Milling machines can produce drilled, reamed and tapped holes in any location.

## Profiles

Profiles are difficult to produce on conventional machines. CNC machines can be programmed to produce complex shapes.

### Indexed patterns

Where workpieces need to be rotated, an indexing device is used. This can create hexagons, octagons and other polygons. A rotary table can be used where arcs or pitch circle diameters are required.

# Features of the workpiece for drilling and grinding

## Drilling

The features of a drilling machine include:

- drilled holes;
- counterbores and countersinks.

### Drilled holes

Holes that go all the way through the workpiece are known as 'through holes' and holes that only go part of the way through the workpiece are known as 'blind holes'.

Usually, the tip of the drill has an angled tip known as a cone angle. This cone angle varies with material but will leave a cone shape at the bottom of blind holes.

Twist drills can be ground flat to create a flat-bottomed hole (see counterbores below).

**Features of drilling**

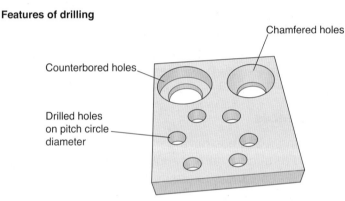

Chamfered holes

Counterbored holes

Drilled holes on pitch circle diameter

**Figure 14.33** Features created by the drilling process

### Counterbores and countersinks

Counterbores produce a hole bigger than the original drilled hole. This is often used to create a recess for the heads of bolts.

A countersink creates a small taper at the entrance to a drilled hole. This can prevent the edge of the hole becoming damaged and can also remove any sharp edges from a hole.

## Grinding

There are two main grinding machines (surface grinders and cylindrical grinders), so we will look separately at the features created by them.

## Surface grinders

These machines are very accurate and the main feature is that workpieces are, therefore, very accurate. They produce very flat surfaces and sides of workpieces at 90 degrees. Steps and shoulders are produced accurately.

Surface grinding features
Radii
Steps
Flat surfaces
Edges

**Figure 14.34** Surface grinding features

## Cylindrical grinders

These machines are very accurate. They produce very accurate diameters and lengths, especially for components which are moving parts. If these components are not accurate the machines they go into may seize. They produce cylindrical shapes, which can be parallel or tapered, and they can be on the outside of the workpiece or the inside of the workpiece, such as a bore. The grinding wheel can be cut to a shape profile, such as a chamfer or radius, which is then ground onto the workpiece.

Threads can be cut using cylindrical grinders. These can be vee-shaped thread, square or other threads. They can produce components with complicated two- or three-start threads and left- or right-hand threads, and they can produce these types of threads externally or internally.

**Figure 14.35** Square thread

# Checks for accuracy using measuring equipment

P5  M1

All the secondary machining techniques that we have been considering (turning, milling, drilling and grinding) need the components that they produce to be checked for accuracy. Measuring equipment and measuring systems are used, although the different systems can differ in accuracy and precision.

## Burrs

These are tiny, sharp spikes of material that are left on the edges of machined components. They cause inaccuracy and can cause scratches to hands.

**Figure 14.36** Measuring equipment

These are checked by eye and, in most cases, you can gently feel for them with a fingernail. Use a small hand file to remove burrs.

## False tool cuts

The cutting tool is designed to produce a machined surface. The quality of the machined surface should be inspected by eye. Often the surface will have marks, grooves or scratches that are not supposed to be there. These are known as false tool cuts.

Typical causes of false tool cuts include:

- blunt tools;
- loose tools or machine parts;
- vibration of the machine;
- workpiece not secured properly.

Where false tool cuts are found, it is best to try to find the cause. If the cause is not found, then the problem may recur.

## Measuring equipment

The following measuring tools are used to measure components on all types of secondary machining techniques:

- steel rule;
- micrometer;
- calipers and vernier calipers;
- vernier protractor;
- dial gauge.

### Steel rule

These are very commonly used and are accurate to 0.5 mm. They are good for checking approximate dimensions but not accurate enough to measure machined components.

### Micrometer

High-precision measuring instruments used for round bar and rectangular shapes. They have an accuracy of 0.01 mm. Some types have an accuracy of 0.001 mm. A lot of skill and practice is needed to use micrometers of this accuracy.

*Outside micrometers* are used for outside diameters. These are manufactured in set sizes. These sizes are 0 mm to 25 mm, 25 mm to 50 mm, 50 mm to 75 mm, and so on.

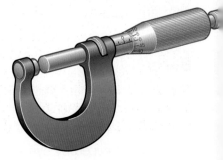

**Figure 14.37** Micrometer

*Inside micrometers* measure bored holes or slots.

*Depth micrometers* can measure the depth of holes or the depth of a step or slot in a machined component.

### Calipers and vernier calipers

Calipers are very versatile and accurate measuring devices. They can measure with an accuracy of 0.01 mm. They can measure lengths, depths, outside diameters and inside diameters.

They are available in sizes from 200 mm to 1 metre long; in special circumstances, they can be even longer. Calipers are particularly useful where a wide range of dimensions needs to be measured; otherwise, a range of micrometers would need to be used.

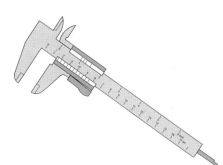

**Figure 14.38** Vernier calipers

Calipers can have different measuring systems, including a dial measurement system, a vernier scale and a digital readout. Machinists have their own personal preferences as to which of these is the best, but they all do the same job.

### Vernier protractor

These instruments are similar to protractors used in maths but are designed to check angles and tapers. On lathes, they typically check tapers and, on milling machines, they check angles.

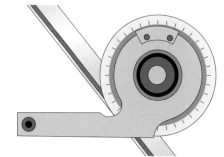

**Figure 14.39** Vernier protractor

### Dial gauge

This is a high-precision instrument. It looks like a watch face but instead of the small deviation reading seconds, it reads distances in 0.001 mm intervals. In some cases, this can vary so that it reads in 0.0001 mm intervals.

**Figure 14.40** Dial gauge

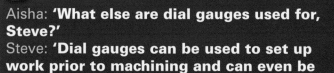

**Team Talk**

Aisha: **'What else are dial gauges used for, Steve?'**
Steve: **'Dial gauges can be used to set up work prior to machining and can even be used to check the accuracy of the machines themselves. They are used on lathes and milling machines to set work in chucks or vices.'**

# Checks for accuracy using measuring systems

P5    M1

In order to manufacture a component to tolerances, it is important to understand how tolerances work. It is impossible to manufacture a component to an exact size, so a size is given with a small amount each side of the dimension. This is known as a tolerance.

For example, a 25 mm diameter shaft can be manufactured 0.5 mm smaller and 0.5 mm larger and still be accepted. This is shown on a drawing as 25.00 +/− 0.5 mm. This gives a total tolerance of 1 mm. When you are machining a component, the tolerance is very important.

## BS4500 limits and fits

An international system is used to help designers give tolerances on components. This is known as BS EN22768-1 or BS4500.

In this system, the tolerances are decided for types of fits. The three types of fits shown here are clearance fits, interference fits and transition fits. Each type of fit has a code based on the size of the hole and the size of the shaft.

*Clearance fits*: the shaft is smaller than the hole, so the shaft will easily go into the hole.

*Interference fits*: the shaft is bigger than the hole, so it will need to be forced into the hole.

*Transition fits*: this is in between clearance and interference fits. Sometimes, there will be a slight clearance fit and sometimes a slight interference fit.

These basic fits are broken down further into codes. Upper case 'H' represents the hole tolerance and a lower case letter represents the shaft.

For example, 'H7 p6' is the code used for bearings to be fitted in to a bearing housing. H7 is the hole tolerance and p6 is the shaft tolerance. The diameter of the shaft determines the amount of tolerance. As the hole gets bigger, the tolerance gets bigger.

### Working to a tolerance of H8

You will need to work to a tolerance of H8, so let's look more closely at that tolerance.

Table 14.3 shows part of the BS4500 system that relates to the H8 tolerance. We will use some examples to determine tolerances.

| H 8 | | | |
|---|---|---|---|
| **Over this size** | **Up to this size** | **Low tolerance** | **High tolerance** |
| 0 mm | 3 mm | 0.000 | 0.014 |
| 3 mm | 6 mm | 0.000 | 0.019 |
| 6 mm | 10 mm | 0.000 | 0.022 |
| **10 mm** | **18 mm** | **0.000** | **0.027** |
| 18 mm | 30 mm | 0.000 | 0.033 |
| 30 mm | 50 mm | 0.000 | 0.039 |

**Table 14.3** H8 tolerance

## Worked Example

**What is the tolerance for a 15 mm H8 hole?**

Solution:

15 mm is bigger than 10 mm but smaller than 18 mm, so we can identify the correct horizontal line to work from (highlighted in Table 14.3). The minimum hole size is 15 mm with a low tolerance of 0.000 mm, so the minimum hole size is 15.000 mm. The maximum hole size is 15.000 mm, plus a high tolerance of 0.027 mm, giving a total of 15.027 mm.

Therefore, the tolerance for a 15 mm H8 hole is 15.000 mm to 15.027 mm.

## Surface texture

Surface texture is a measure of the roughness or smoothness of a surface of material. If we magnified the surface of a machine's workpiece, it would be rough. It would have tiny peaks and valleys, like a mountain range but on a microscopic scale.

A surface measurement system known as 'roughness average' (Ra) is used to give the roughness of the surface a value. This value is the average distance from a centre line of the peaks and valleys to the height or depth of the peaks and valleys.

Each machining process has the ability to produce a level of surface finish. Table 14.4 below shows a range of processes. It also shows the typical types of finish that each process can produce. From the table, we can see that milling and turning can produce work from Ra 1.0 micrometers to around 8.0 micrometers (this is shown by the green bar). The yellow bar indicates that, under exceptional circumstances, a better or worse finish can be created. Note that an extra process of polishing is shown – this is to highlight that no process can give a perfect finish.

To achieve P5 in this unit, you need to work to the following surface finishes:

- reaming – 1.6 µm (63 µin);
- turning – 1.6 µm (63 µin);
- milling – 1.6 µm (63 µin);
- grinding – 0.2 µm (8 µin).

| | Roughness Average value (Ra) | | | | | | | | | | |
|---|---|---|---|---|---|---|---|---|---|---|---|
| *Micro inches* | **1000** | **500** | **250** | **125** | **63** | **32** | **16** | **8** | **4** | **2** | **1** |
| *Micro meters* | 25.4 | 12.7 | 6.4 | 3.2 | 1.6 | 0.8 | 0.4 | 0.20 | 0.10 | 0.05 | 0.03 |
| Drilling | | | | | | | | | | | |
| Reaming | | | | | | | | | | | |
| Milling & Turning | | | | | | | | | | | |
| Grinding | | | | | | | | | | | |
| Polishing | | | | | | | | | | | |

**Table 14.4** Roughness average values

## Drilling and reaming

If you look at Table 14.4, you can see that it is difficult to drill a hole to 1.6 µm as it is beyond the capability of the machine. If a reamer is used, it can easily produce the quality of finish needed. The correct speed and feed should be used, as should coolant.

## Milling and turning

For milling and turning, a surface finish of 1.6 µm is needed. This is almost at the limit of the machine's capability. For this reason, you should consider the following:

- The tool should be sharp and in good condition.
- The workpiece should be held securely.
- The correct speeds and feeds must be used.
- The machine should be bolted to the floor.
- Coolant should be applied.
- Small finishing cuts should be taken.

## Grinding

You will need to grind to a Ra of 0.2 μm. You can see from Table 14.4 that this is almost at the maximum capability of this process. The table also shows that grinding can give a poor finish, so getting the parameters right is especially important with grinding machines. You will need to consider the following:

- The wheel should be dressed with a diamond.
- The workpiece should be securely clamped.
- The feed should be slow.
- The coolant should be on full power.
- Small cuts of 0.003 should be applied on the finishing cut.
- The wheel should pass over the surface two or three times on the final cut.

### Make the grade

The next activity will help you in achieving the following grading criterion:

 **P5** machine a given workpiece safely and carry out necessary checks for accuracy.

### Activity

Use the component product shown in Figure 14.29 for criterion P4 on page 288. You should produce the component.

Complete a table showing all safety equipment and measuring equipment used.

- Use appropriate measuring equipment.
- Use appropriate safety equipment.

Table 14.5 shows a completed table.

| Checking workpiece accuracy | | | | Witness signatures |
|---|---|---|---|---|
| **Safety equipment** Wear boots, overalls and goggles. | | | | Signature |
| **Safety practice** Work with supervision. Ensure guards are in place. Ensure work area tidy. | | | | Signature |
| | **Element** | **Equipment** | **Outcome** | **Action** | **Student checks work** |
| 1 | 30mm+/- 0.02 | 25–50 micrometer | 30.01 mm | In tolerance | Signature |
| 2 | Slot 50mm +/- 0.1 | 150mm vernier caliper | 50.08mm | In tolerance | Signature |
| 3 | Depth 11.50mm +/- 0.05 | 0–25 depth micrometer | 11.47mm | In tolerance | Signature |

**Table 14.5** Checking workpiece accuracy

## Activity

Explain why it is important to carry out checks for accuracy of features on components during and after manufacture.

## Make the grade

This activity will also help you in achieving the following grading criterion:

**M1** explain why it is important to carry out checks for accuracy of features on components during and after manufacture.

# Learning Outcome 4. Know about aspects of health and safety relative to secondary machining techniques

# Health and safety

This section considers how the law relates to secondary machining techniques and includes information on:

- UK legislation;
- European directives;
- reducing risks:
  1. risk assessment;
  2. avoidance of dangerous conditions.

# UK legislation

In the United Kingdom, the main piece of legislation relating to occupational health is the Health and Safety at Work Act 1974 (HASAW).

This act is broken down into parts and sections, and many Regulations follow it. Regulations relate to specific areas for all types of working situations. To ensure that businesses work to this legislation, the Health and Safety Executive (HSE) enforces the law and investigates any major issues.

For further information on this refer to Unit 1 Working safely and effectively in engineering

# European directives

These are laws set by the European Union (EU). They require the member countries of the EU to achieve certain results, but they do not dictate the means of achieving those results.

The HSE website (www.hse.gov.uk) gives detailed information on the regulations relating to secondary machining processes.

Some important regulations that relate to secondary machining are outlined below:

- *Safety in the use of abrasive wheels:* this guidance was revised in line with the Provision and Use of Work Equipment Regulations 1998 (PUWER 98).
- *Control of Substances Hazardous to Health (COSHH):* see Team Talk below.
- *BS EN12840:2001:* safety of machine tools – manually controlled turning machines with or without automatic control.
- *BS EN12717:2001:* machine tools – safety – drilling machines.
- *BS EN13218: 2002:* machine tools – safety – stationary grinding machines.

# Reducing risks

### Risk assessment

A risk assessment is an examination of what could cause harm to people in a workplace. A risk assessment will help

> **❝ Team Talk**
>
> Aisha: **'What do the initials COSHH stand for, Steve?'**
> Steve: **'Control of Substances Hazardous to Health. This legislation requires employers to control any substances that can harm workers' health.'** ❞

a business decide whether enough precautions have been taken to prevent accidents. Businesses are legally required to assess the risks in their workplace so that a plan can be made to control any risks.

There are five steps to making a risk assessment:

*Step 1* Identify any hazards that may be present.
*Step 2* Decide who could be harmed by the hazards and how they could be harmed.
*Step 3* Evaluate the risks that have been found and decide on the precautions needed.
*Step 4* Record all findings and then implement them.
*Step 5* Review the assessments and update if necessary.

### Avoidance of dangerous conditions

Although there are laws, legislation, European directives and risk assessments, there is always the chance of dangerous occurrences in the workplace. If you see anything that is dangerous or that could be dangerous, tell your supervisor. They can assess whether there is a risk, either from experience or with the support of a safety advisor.

Dangerous conditions can be caused by people, so poor behaviour and dangerous working methods should also be considered.

## Working safely when using secondary machining techniques

All secondary machining processes are potentially dangerous. The following aspects must be considered:

- moving parts;
- machine guards;
- handling cutting fluids;
- insecure components;
- emergency stops;
- machine isolation;
- personal protective equipment (PPE);
- clean and tidy work areas.

### Moving parts

All moving parts should be guarded.

## Machine guards

Machine guards protect operators from a number of problems, such as tools breaking and workpieces coming loose from the work-holding device. Machine guards prevent fast-moving parts hitting the operator. They also prevent the operator's clothing becoming caught and potentially pulling the operator into the machine.

## Handling cutting fluids

Cutting fluids such as coolant are generally made by mixing water with natural oils. Over a period of time, these coolants lose their capacity to cool properly. As they are natural products, they will eventually form bacteria within them and failure to change coolants regularly may lead to a build-up of harmful bacteria, which can irritate skin. It is not practical to wear gloves when machining, so a special cream called 'barrier cream' is used. This protects the skin from harmful bacteria.

## Insecure components

If a workpiece, tool or other component is insecure, it may be ejected from the machine tool very quickly and dangerously. Care must always be taken to ensure components are fastened securely. Often nuts and bolts holding components in place can vibrate loose over a period of time. These must be checked regularly.

## Emergency stops

Around a workshop, you will see emergency stop buttons. These are large red, domed buttons. They are designed so they can be pressed easily from anywhere in a workshop in the event of an accident. If you believe that there is an emergency situation in the workshop, you should immediately press one of these buttons. However, do not press the emergency stop unless there is an emergency because, once pressed, every machine in the workshop will be switched off.

## Machine isolation

Every machine has a start button (usually green) and a stop button (usually red). There will also be an isolation switch for each machine. If this is switched to the 'off' position, then no power will reach that particular machine. The isolation switch should be used when setting tools or changing gears.

## Personal protective equipment (PPE)

All operators must wear protective clothing.

- *Goggles:* these protect eyes from material chips, swarf, broken tools and coolant. These should always be worn in the workshop.
- *Overalls:* these protect against coolant splashes and prevent loose clothing and belts becoming caught in machinery.
- *Safety boots:* these have thick soles to protect the soles of the feet from sharp objects and they give good grip in workshops where oil and coolant are often found on the floor. They have steel toecaps, which prevent toes being crushed if a heavy object falls onto them. They are robust and sturdy to protect feet and ankles from sharp objects and material found around the workshop.

## Clean and tidy work areas

See Team Talk below.

### 66 Team Talk

Aisha: **'What do we mean by "housekeeping"?'**
Steve: **'Housekeeping means keeping the work area clean. It involves keeping tools in a safe place, sweeping up any scraps of metal filings or emery cloth. It also includes brushing and wiping down machines after use. Walkways should be kept clear and there should be no materials or objects causing an obstruction or dangerous situation in the work area.'** 99

# Working safely when turning, milling, drilling and grinding

P6    M2    D2

In this section, we will consider working safely in all of the secondary machining techniques.

## Swarf

One problem that is common to the processes of turning, milling and grinding is swarf. Swarf is the name given to material that has been removed from workpieces by machining. Material is cut away from the workpiece in lengths of thin, sharp material. These can be small lengths, known as chips, or long lengths like ribbons. Goggles should be worn to protect the eyes.

There are a number of problems associated with swarf:

- It is usually very hot.
- It is extremely sharp, like a razor blade.
- It can be continuous and move in an unpredictable way.
- It can be caught up in the cutter and explode in the machine area.
- It can become entangled with clothing.

Swarf should be removed after machining is finished. Special tools with hooks are used to remove swarf. **Do not** use bare hands to remove swarf.

Always use hand brushes to remove swarf from worktables.

On a lathe, the swarf is held in the tray of the machine, away from the work area, but it can build up easily. This must be removed if it builds up to high levels.

## Turning safely

### Handling tools

Turning tools have two distinct parts: the shank of the tool, which is often painted in a bright colour, and the tip of the tool, which is a shiny, silver metal. The shank is often cast and is not sharp, so the tools should be carried by this part. The tip is extremely sharp and should not be touched. Following a machining operation, the tip can be extremely hot, adding to the danger.

### Tool breakages

A turning tool can easily break during the cutting process. The tip can break away. Goggles should be worn to protect eyes.

Parting tools are long and slim, and can easily break. When they do break, there will be a bang. A large piece of tool can be very dangerous. Guards are used to protect the operator from this type of break.

## Milling safely

Milling tools are extremely sharp and, for this reason, are stored in individual boxes when not in use. Vertical milling cutters such as end mills and slot drills have a plain shank that does not have cutting edges, so they should be held by this section.

Horizontal milling cutters such as slab mills have no shank and should be handled with extreme care. For extra safety, a cloth can be held while the cutter is lifted.

When slab mills break, large sections of cutting tool can be sent into the direction of the operator. Guards must be in place to protect against this.

In conventional milling machines, a problem known as backlash can occur. This is when the cutter jumps forward slightly, which can cause tools to break. A feature known as a backlash eliminator is used to prevent this.

## Drilling safely

Twist drills have sharp edges but, if handled carefully, do not cause injury. Twist drills are changed relatively often. As with all cutting tools, they can become hot when in use, so when handling these tools after an operation, care should be taken.

As twist drills become smaller, the speed at which they spin becomes faster. If a drill breaks at high speed, pieces of the drill will be thrown out of the machine. Guards and goggles are there to protect the operator from this.

## Grinding safely

Grinding is potentially the most dangerous of all of the secondary machining techniques because the ceramic wheel revolves at high speed. In a worst-case scenario, the wheel can explode, known as wheel bursting.

Extreme care is needed when setting and balancing the wheel. Supervision is needed when using grinding wheels. Do not attempt to use a grinding machine unless you have been specifically told to do so.

Grinding wheels should be kept in individual boxes and stored safely. If these wheels are dropped, small fractures can lead to disastrous outcomes.

### Sparks and airborne particles

In the process of grinding, sparks are created. This is the metal heating up as it is ground from the workpiece. Minute particles of metal are expelled at high speed and at red-hot temperatures. They can be quickly cooled using coolant, but care should be taken to avoid contact with these airborne particles.

## Make the grade

This activity will help you in achieving the following grading criterion:

**P6**    **describe methods of reducing risk for the secondary machining technique used.**

## Activity

**Describe five general methods of reducing risks when using lathes, milling machines, drilling machines and grinding machines.**

**Give two specific methods of reducing risk for each of these machines.**

## Make the grade

This activity will help you in achieving the following grading criterion:

**M2**    **explain the importance of using the correct tooling and having machine parameters set correctly when machining a workpiece.**

## Activity

**Explain the importance of using the correct tool during a machining process. Consider the effects of accuracy of the workpiece and the safety problems that may occur when incorrect tools are used.**

**Explain how incorrect speeds, feeds and cut sizes can impact upon accuracy of the workpiece and how this can cause safety issues.**

## Make the grade

The next activity will help you in achieving the following grading criterion:

**D2** compare and contrast three secondary machining techniques for accuracy and safety of operation.

## Activity

1. Compare and contrast the accuracy of the following machines:
   - pedestal drill;
   - centre lathe;
   - surface grinder.
2. Compare and contrast the safety of operation of these machines.

 **Grading criteria recap**

To achieve a **pass grade** you must be able to:

**P1** describe how three different secondary machining techniques are used;

**P2** describe the appropriate use of three different work-holding devices for these different techniques;

**P3** describe the appropriate use of three different tools for these different techniques;

**P4** monitor and adjust the machining parameters to machine a given workpiece correctly and safely and to produce features as defined by the workpiece;

**P5** machine a given workpiece safely and carry out necessary checks for accuracy;

**P6** describe methods of reducing risk for the secondary machining technique used.

To achieve a **merit grade** you must be able to:

**M1** explain why it is important to carry out checks for accuracy of features on components during and after manufacture;

**M2** explain the importance of using the correct tooling and having machine parameters set correctly when machining a workpiece.

To achieve a **distinction grade** you must be able to:

**D1** justify the choice of a secondary machining technique for a given workpiece;

**D2** compare and contrast three secondary machining techniques for accuracy and safety of operation.

# Unit 18
## Engineering marking out

# Introduction to the unit

Engineers, manufacturers and machinists who make engineered parts and components that are machined or cut out using hand processes need to 'mark out' the workpiece before processing. 'Marking out' refers to the process of transferring the information contained on an engineering drawing to the piece of material that is to be machined.

This unit looks at the special equipment and tools that are needed to mark out workpieces. It will help you understand the correct equipment to use and how to use it, and provide a guide on how to work safely and store equipment correctly.

This unit will also provide guidance on how to plan your marking out and further support your understanding of interpreting engineering information as the basis for marking out.

**Learning Outcomes**

By the end of this unit you should:

- know about marking-out methods and equipment for different applications;
- be able to mark out engineering workpieces to specification.

# Grading criteria

| To achieve a pass grade you must be able to: | To achieve a merit grade you must be able to: | To achieve a distinction grade you must be able to: |
| --- | --- | --- |
| **P1** select suitable measuring and marking-out methods and equipment for three different applications | **M1** recommend corrective action for unsafe or defective marking-out equipment | **D1** justify the choices of datum, work-holding equipment and measurement techniques used to mark out the three different applications |
| **P2** describe the measuring and marking-out equipment used for the three different applications | **M2** carry out checks to ensure that the marked-out components meet the requirements of the drawing or job description | |
| **P3** prepare a work plan for marking out each of the three different applications | | |
| **P4** mark out the three different applications to the prepared work plan | | |
| **P5** demonstrate safe working practices and good housekeeping | | |

# Learning Outcome 1. Know about marking-out methods and equipment for different applications

## Work-holding devices

Before we go any further, you will need to remember and understand a few important terms that are used in this unit.

> ### Key words
>
> **Engineering drawing** – this is the drawing that is used to technically describe a product. It is drawn using methods that are standardised throughout the world so that engineers work to the same basic rules of presenting information.
>
> **Workpiece** – this is the engineering material (often a metal or polymer) that will be cut or machined to make the final product.
>
> **Product** – this is the final manufactured part or component.

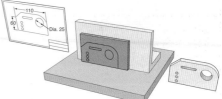

## Holding the workpiece in preparation for marking out

If we were drawing on paper, we would be drawing on a flat surface, so everything would be drawn in two dimensions: length and height. When we mark out, we have to transfer the 2D drawing to a workpiece in three dimensions: length, height and depth. Therefore, we need to consider how to hold the workpiece and align it in the right position to help us mark out effectively.

**Figure 18.1** An engineering drawing and completed product

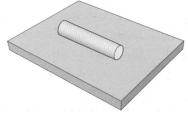

Figures 18.2 and 18.3 show two workpieces. It would be difficult to mark them out as shown because they may move about, thereby making the process highly inaccurate.

**Figure 18.2** Workpiece 1

In this section, we will introduce the types of equipment known as holding devices, which keep workpieces firmly in place.

In the process of marking out, all work surfaces and work-holding devices should be clean and a high degree

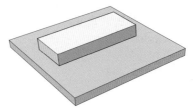

**Figure 18.3** Workpiece 2

of care should be taken when using them. Accuracy of marking out often depends upon the cleanness and quality of the work-holding equipment.

There are a number of common work-holding devices, as listed below:

- vee blocks;
- angle plates;
- clamps;
- parallels;
- toolmaker's vice.

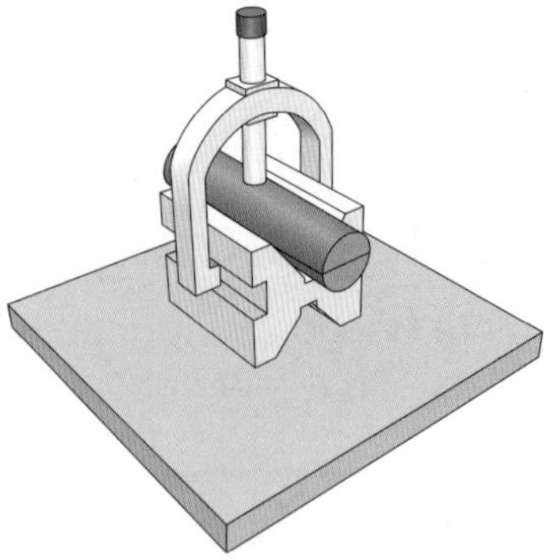

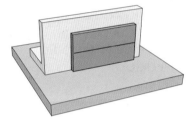

**Figure 18.4** Work-holding device

Some of these devices are also described in Unit 14, as they are used to hold and clamp workpieces in preparation for machining.

## 66 Team Talk

Aisha: **'Why is it important to clean and store equipment?'**
Steve: **'Keeping marking-out equipment clean and carefully storing it when finished will provide better accuracy when marking out future workpieces. It will also keep the equipment in better condition for longer.'**
Steve: **'The other reason is that it will be easy to find the equipment next time.'** 99

**Figure 18.5** Angle plate

### Vee blocks

A vee block is a square block of material with an accurate 'V' shape machined into it. This is very accurate and is often used to hold round bar, preventing the bar from rolling out of position. The bar is often secured with a clamp.

### Angle plates

Angle plates are 'L' shaped and are manufactured to exactly 90 degrees. When a workpiece is aligned on an angle plate, it is exactly perpendicular (90 degrees) to the work surface.

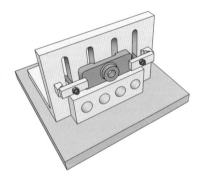

**Figure 18.6** Clamp

### Clamps

Clamps are often used in conjunction with an angle plate to keep the workpiece in position. Clamps are tightened

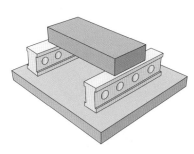

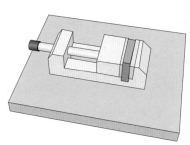

**Figure 18.8** Toolmaker's vice

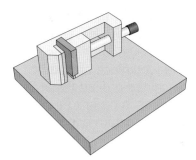

**Figure 18.9** Rotated toolmaker's vice

**Figure 18.7** Parallels

with a spanner (or wrench) and stepped clamping blocks are used at the back end of the clamp. Care must be taken not to damage the workpiece by over-tightening the clamp because soft materials mark very easily.

## Parallels

Parallels are flat, accurate rectangular bars. They often come as a pair and can be used to lift the workpiece from the surface plate. They should be cleaned after use and stored in a case or designated storage area.

## Toolmaker's vice

A toolmaker's vice is designed so that all sides are aligned at exactly 90 degrees to each other. It can be rotated through 90 degrees in all directions, which saves time because the workpiece does not need to be removed and reset.

In Figure 18.9, the toolmaker's vice has been rotated quickly and effectively to allow marking out on a different workpiece face.

# Datums and referencing

## Creating a datum

Figure 18.10 shows an engineering drawing and a workpiece. A red point has been used to show a reference point known as a datum. All measurements are taken from this point. The person marking out will be given an engineering drawing, but the datum point may not be shown on the drawing. They will need to decide the best

> **Key word**
>
> A **datum** is a point from which all dimensions are taken.

place to put the datum. A good position is a corner with a 90-degree edge. The workpiece should be machined or filed flat at 90 degrees in order to establish the two datum edges. When the workpiece is marked out, it will be located on one of these datum edges.

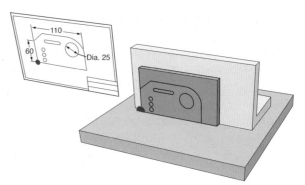

**Figure 18.10** Creating a datum

## Accuracy of datum edges

When marking out a workpiece, it is important to know where all the measurements will be taken from and that the point from which all measurements are taken is flat and smooth.

In Figure 18.11, a measurement from the left-hand side of the workpiece shows 57 mm on the bottom edge and 56 mm on the top edge. This means the workpiece will have errors, which could cause quality problems.

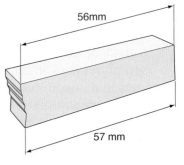

**Figure 18.11** Accuracy of datum edges

## Setting of datums on circular components

In Figure 18.12, there are no square edges on the workpiece. In this product, the centre of the component can be used as a datum. The red spot indicates the centre of the drawing. The details and sizes have been transferred accurately onto the workpiece.

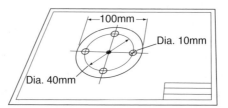
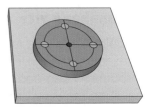

**Figure 18.12** Setting datums on circular components

## Marking blue

Marking blue is a special ink used in metalworking and marking out. It is used to coat a workpiece with a very thin layer of ink. This ink can then be scratched off (with a scribe or other such tool) to reveal a bright, narrow line as the metal underneath is revealed.

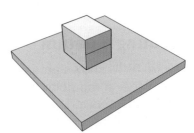

**Figure 18.13** Marking blue

**❝ Team Talk**

Steve: **'Before marking out, take some time to make sure you have chosen the best position for the datum on your workpiece.'** ❞

# Basic marking-out equipment

**Remember:** marking out simply refers to a drawing produced on the metal or material. If you were to draw a shape on paper and then cut the shape out, you would use tools such as a pen or pencil and a rule or protractor. If we tried to use these tools on metal, the pen or pencil would not show clearly enough on the metal to be accurate. This section will help you recognise and choose the special type of equipment needed to mark out the material.

The basic marking-out tools are as follows:

- steel rule;
- scriber;
- centre punch;
- engineer's square.

## Steel rule

This is a basic measuring and marking-out tool. It differs from a rule used for drawing on paper in several ways. It is made from thin sheet steel with the graduations (or increments) starting at the very edge of the rule (rather than a few millimetres in, as on a standard rule).

Steel rules are available in a range of sizes, including 150 mm, 300 mm, 600 mm and 1,000 mm (1 metre). Graduations are in 0.5 mm and 1 mm. This is a relatively inaccurate tool but useful for reading a dimension or size to an accuracy of within half a millimetre.

**Figure 18.14** Steel rule

## Scriber

A scriber is a long, steel tool with sharp points, which are designed to mark a line on the metal. This is known as 'scribing' a line. These tools will scribe a clear line straight onto a soft material workpiece (such as aluminium) or through the use of marking blue on a harder material (such as medium-carbon steel). It is important to keep the points as sharp as possible as this will provide a more accurate line (similar to a sharp pencil when drawing).

**Figure 18.15** Scriber

**Team Talk**

Steve: **'You have to be careful when using a scriber.'**
Aisha: **'Why?'**
Steve: **'A scriber has sharp points to create an accurate line on the material; they can be single or double ended. You have to be careful of these sharp points as they can easily scratch the surface of your skin.'**

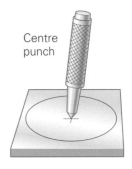

Centre punch

Figure 18.16 Centre punch

## Centre punch

A centre punch is designed to create small indents in the metal by using a force applied with a hammer.

These indents are used for a number of reasons, as outlined below.

1. Locating the position of a hole that will be drilled. The indent helps the drill locate the correct position.
2. Creating a small indent to allow the use of dividers.
3. Creating a series of indents along a profile, which is slightly more visible than a purely scribed line. If the marking blue is removed, the small indents will still be clearly visible.

## Engineer's square

Figure 18.17 Engineer's square

An engineer's square is used to produce a perpendicular line to an edge; in other words, a line 90 degrees to one edge of the material.

It is an L-shaped tool that is very accurate and manufactured to 90 degrees. The heavy end is pressed against the edge of the metal and the thin, steel end is rested on the workpiece at exactly 90 degrees to the edge. A scriber is used to create a line by running the point along the thin end of the engineer's square.

Engineer's squares are also used to inspect or check workpieces already produced. The engineer's square is laid over the workpiece to check whether it is at 90 degrees.

### Make the grade

This activity will help you in achieving the following grading criterion:

**M1** recommend corrective action for unsafe or defective marking-out equipment.

## Activity

**Using a range of marking-out equipment, complete Table 18.1.**

## Marking out – corrective action

| Equipment | Tick (good) or cross (not good) | | | | | Student's comments | |
| | Visible graduations | Sharp points | Clearness of surfaces | Dent-free surfaces | Moveable screws | Other problems | Actions |
|---|---|---|---|---|---|---|---|
| Steel rule | | N/A | | | N/A | | |
| Scriber | N/A | | | | N/A | | |
| Centre punch | N/A | | | | N/A | | |
| Engineer's square | N/A | N/A | | | N/A | | |
| Dividers | N/A | | | | | | |
| Odd-leg calipers | N/A | | | | N/A | | |
| Scribing block | N/A | | | | | | |
| Surface tables/ plates | N/A | N/A | | | N/A | | |
| Vernier protractor | | | | | | | |
| Vernier height gauge | | | | | | | |

**Table 18.1** Marking out – corrective action

Corrective actions may include cleaning graduation scale, sharpening tool points or loosening moveable screws. Other problems may include missing components or severely damaged parts.

# Further marking out equipment

P1   P2   M1

This section will introduce a number of very specific tools that are used in marking out, as follows:

- dividers;
- odd-leg calipers;
- scribing block;
- surface table/plate;
- vernier protractor;
- vernier height gauge.

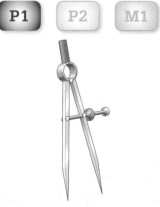

**Figure 18.18** Dividers

## Dividers

Dividers are used to produce circles or arcs on the material. They are used in the same way as a pair of compasses on a piece of paper. Instead of one point and one pencil, there are two points, known as 'legs'. Essentially, it is like two scribers hinged together. One point is located into an indented hole produced by a centre punch and the other point is used to create the circle or radius. A small screw is adjusted to move the legs apart and a steel rule is used to measure the distance between the points. They are accurate to within 0.5 mm.

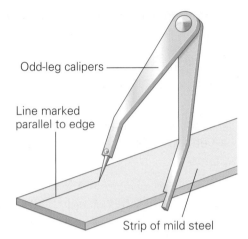

Odd-leg calipers

Line marked parallel to edge

Strip of mild steel

**Figure 18.19** Odd-leg calipers

## Odd-leg calipers

Odd-leg calipers, often referred to as just 'odd-legs' are used to scribe lines parallel to a straight edge. One leg has a square cut into it which sits on the edge of the material. The other leg is a sharp point and is used as a scribe. As the odd-legs are moved along the edge of the metal, the scriber produces a line parallel to this edge. The correct distance is set using a steel rule, moving the legs to suit the distance required.

**Figure 18.20** Scribing block

## Scribing block

A scribing block is a special tool used to draw lines on workpieces that are parallel to a flat surface. A long, thin scribe is used, which is adjustable in all directions using a number of screws that are tightened and fine-adjusted by hand. The bottom surface needs to be clean and perfectly flat, as does the surface on which the scribing block is sitting. The height is set using a steel rule, providing accuracy of about 0.5 mm.

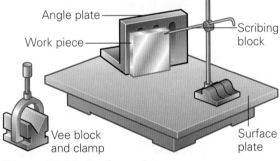
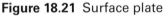

Angle plate

Work piece

Scribing block

Vee block and clamp

Surface plate

**Figure 18.21** Surface plate

## Surface table/plate

Surface tables and surface plates are used to ensure that the surface on which the marking out takes place is perfectly flat. A desktop or workbench is generally not suitable for accurate marking out. Surface plates are smaller and can be placed on benches. Marking-out tables are extremely heavy and cannot be moved easily. It is poor practice to use a surface table as a general workbench – the clean, flat surface can easily become damaged

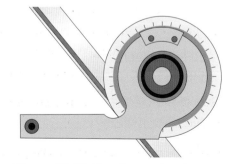

**Figure 18.22** Vernier protractor

by common hand tools such as a hammer or file. Surface tables and plates should always be covered after use to protect them.

## Vernier protractor

A vernier protractor is used to draw angled lines on the workpiece. The angle is created accurately by the vernier scale around the circumference of the protractor.

The handle is precise and flat, and is pressed against the edge of the workpiece, which must also be flat and clean. The blade is now in position, indicating the angle on the workpiece, and a scriber is used to mark a line at the required angle. The accuracy of the angle will depend on how accurately the angle is set on the protractor's vernier scale. The line is also affected by the sharpness of the scriber.

## Vernier height gauge

A vernier height gauge is a tool used to scribe lines that are parallel to the surface table or plate. It is a very accurate device as it combines a vernier scale with a fine-adjustment screw to provide dimensions of 0.02 mm accuracy.

These gauges are available in a number of heights, including 300 mm, 400 mm and 500 mm. Height gauges are also available with digital readouts, which provide a quick and easy method of setting and reading.

## Vernier scale

A vernier scale is an accurate method of measuring. It has a main scale like a rule which shows sizes to within 1 mm as well as a secondary scale which shows sizes to an accuracy of 0.01 mm.

**Figure 18.23** Vernier height gauge

### Make the grade

This activity will help you in achieving the following grading criteria:

**P1** select suitable measuring and marking-out methods and equipment for three different applications;

**P2** describe the measuring and marking out equipment used for the three different applications.

## Activity

The three products A, B, and C, shown in Figure 18.24, are to be marked out. For simplicity, all dimensions are missing – you will still be able to select and describe appropriate equipment without the need of dimensions.

Select and describe measuring and marking-out equipment for the three products, using the following headings:

- Select measuring and marking-out equipment
- Describe measuring and marking-out equipment.

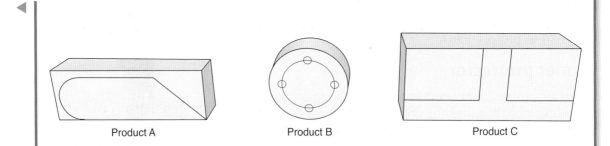

**Figure 18.24** Products A, B and C
**An example of what is needed is given for Product A.**

**Product A:**
**Select measuring and marking-out equipment.**

*Vernier height gauge*
*Hammer*
*Centre punch*
*Vernier protractor*
*Scriber*
*Steel rule*
*Dividers*

**Describe measuring and marking-out equipment.**

**Vernier height gauge**
A tall measuring device with a vernier scale. This can be set to any height. It has a sharp tool which can scribe a horizontal line on workpieces.

**Hammer**
This has a metal head and wooden handle. It is used to hit a centre punch to create indents for dividers.

**Centre punch**
A long, thick solid steel tool shaped like a pen. It has a sharp point  so when it is hit with a hammer it causes an indent into the workpiece.

**Vernier protractor**
This is an accurate device for measuring angles. It has one straight edge to sit on the workpiece and one straight edge which a scriber is moved along to create a line on the workpiece.

**Scriber**
A long, thin solid steel tool shaped like a pen with a sharp point. It is held in the hand and is used to scribe lines on a workpiece.

◀ **Steel rule**

A long, thin and flat measuring device. It has divisions of 0.5 mm. Used for measuring a workpiece and setting dividers.

**Dividers**

Shaped like a drawing compass, it has two legs both with sharp points. One point sits in an indented hole, the other leg scribes an arc or circle.

# Advanced marking-out equipment

Advanced marking-out equipment relates to devices that are more complex or initially can be difficult to use. Here we look at:

- dial test indicators (DTIs) or dial gauges;
- slip gauges (or gauge blocks);
- angle plates;
- vee blocks and clamps.

## Dial test indicators (DTIs) or dial gauges

A dial test indicator (DTI) is used to check work or set work in a machine tool such as a lathe.

It is extremely accurate and can measure workpieces within 0.001 mm (a thousandth of a millimetre, or a micron). These DTIs are held on a scribing block or on a stand with a magnetic base.

## Slip gauges or gauge blocks

Slip gauges are small blocks that are extremely accurate. They range from 0.5 mm to 100 mm in a standard workshop set. The blocks are placed together and are held in place through air pressure. Blocks are joined or 'wrung' together and are built up to the size required. To prevent the manufacture of very thin gauges, the smallest blocks are around 1 mm thick. The sizes increase in 0.001 mm, e.g. 1.000 mm, 1.001 mm, 1.002 mm, and so on.

To illustrate, if we needed to measure a dimension of 142.34 mm, we could use the following 'slips':

- 100 mm;
- 40 mm;

### Key word

**Wrung** – pushing two slip gauges together and twisting. This removes all air between them and makes them stick together.

**Figure 18.25** Dial test indicator

**Figure 18.26** Slip gauges

- 1.3 mm;
- 1.04 mm.

Total: 142.34 mm

## Angle plates

Angle plates are L-shaped cast-iron or steel devices that help ensure that workpieces are held at 90 degrees to the surface plate or surface table. They often have slots in them which allow workpieces to be bolted to them while also making the plate lighter. These devices can be very heavy, so always check with the workshop supervisor before lifting.

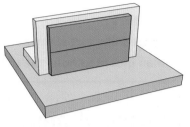

**Figure 18.27** Angle plates

## Vee blocks and clamps

A vee block and clamp is a special device used to hold round bars. It is very accurate and should be wiped clean before use. The round bar sits in the vee-shaped cut and a small clamp is hand-tightened onto the workpiece.

Vee blocks are available in a range of sizes, so do not try to use one that is too small or too big. They are used to mark centre lines on round bars; this can be along the side of the bar or on the end of the bar, known as the face.

## Calibration

After time, marking-out equipment such as vernier height gauges can become inaccurate. The process of adjusting equipment so it gives an accurate reading is known as calibration.

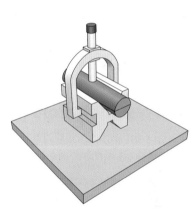

**Figure 18.28** Vee block and clamp

In workshops, gauge blocks are often used for calibration. If a gauge block of 50 mm height is used to check a height gauge, then the height gauge should read 50 mm exactly. If it does not, it can be adjusted using small screws so that it reads exactly 50 mm.

From time to time, even the gauge blocks must be sent away to a specialist to be checked.

When both instruments have been calibrated against a standard, their readings will then be against the standard.

## Marking-out mediums

Marking blue is most commonly used in manufacturing workshops. It is quick and easy to use but is best for small-sized areas. Where large areas of material are used, such as fabrication and welding, whitewash can be used.

---

### Key words

**Precision** – When an instrument gives very similar readings when measuring a workpiece a number of times this is precise.

**Accuracy** – This is when the instrument gives a reading very near to or exactly the true size of the workpiece.

This is a liquid or solid chalk. The chalk is not actually chalk but a small block of soapstone that can be sharpened to a point. This 'chalk' is white, so it stands out well on metal, which is often grey and dirty or has some corrosion.

## Make the grade

The next activity will help you in achieving the following grading criterion:

**P3**  **prepare a work plan for marking out each of the three different applications.**

| **Marking out – work plan** | | | | |
|---|---|---|---|---|
| **Product name Product A** | | | | |
| Operation | Marking-out tool | Work-holding devices | Description of process | Safety |
| Parallel lines | Vernier height gauge | Use marking-out table (hand-held). | Hold workpiece flat to the table. Set the vernier height gauge and create top line and centre line parallel to the table. Rotate workpiece 90 degrees. Set height and mark arc centre line. | Be careful of sharp edges of height gauge. Replace height gauge when finished. |
| Create angle | Vernier protractor Scriber | Use marking-out table. | Set vernier protractor and scribe line. | Be careful of sharp edges. |
| Create arc | Centre punch Hammer Dividers Stamping block | Use marking-out table. Use stamping block when hammer is used. | Centre punch a small indent at the arc centre. The workpiece must be put onto a stamping block to do this. | Be careful when using hammers. Wear goggles. Do not use marking-out table when creating centre-punch indent. |

**Table 18.2** Marking out work plan for Product A

## Activity

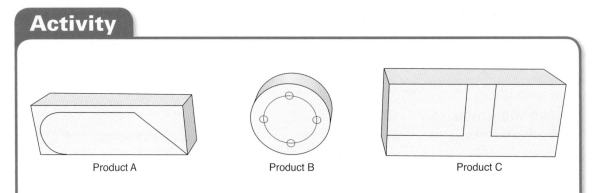

Product A          Product B          Product C

**Figure 18.29** Products A, B and C

**Prepare a marking-out work plan for each of these products. An example work plan is shown in Table 18.2. This would be suitable for Product A.**

**Fill in Table 18.3 for Products B and C.**

| Marking out – work plan | | | | |
|---|---|---|---|---|
| Product name | | | | |
| Operation | Marking-out tool | Work-holding devices | Description of process | Safety |
| | | | | |
| | | | | |
| | | | | |
| | | | | |

**Table 18.3** Marking out work plan

# Marking out regular-shaped workpieces

In this section, we will consider the different marking-out devices and tools used to mark out workpieces of various forms. We will look at:

- cube- or block-shaped metal;
- flat plate;
- holding thin plate vertically;
- round or cylindrical bar;
- discs.

Figure 18.30 illustrates the key features we need to consider to mark out a workpiece effectively: the surface plate, the workpiece, the part, the application of marking blue and the subsequent scribed line.

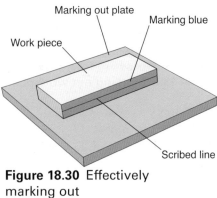

Marking out plate
Marking blue
Work piece
Scribed line

**Figure 18.30** Effectively marking out

## Cube- or block-shaped metal

When a material has a flat surface, it is perfectly feasible to lay it on the marking surface and hold the workpiece by hand. A scribing block or vernier height gauge can then be used to scribe the line as shown in Figure 18.31.

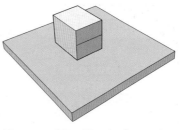

**Figure 18.31** Block-shaped metal

## Flat plate

As with a cube- or block-shaped metal, a flat plate is quite stable and can be secured manually.

## Holding thin plate vertically

Thin plate held vertically will move backwards and forwards, which will result in an inaccurately scribed line. An angle plate can be used to ensure that the plate is exactly 90 degrees to the surface plate. This can be held by hand or in some cases it may be clamped to the angle plate.

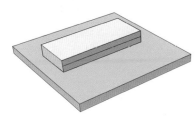

**Figure 18.32** Flat plate

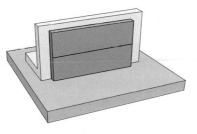

**Figure 18.33** Holding thin plate vertically

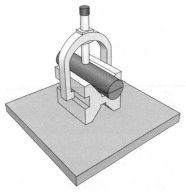

**Figure 18.34** Clamping a round bar

## Round bar

Round bar can be difficult to hold while simultaneously marking out. A vee block and clamp will ensure that the bar is perfectly parallel to the surface plate. The bar can then be marked along its side or on the face.

### Marking out lengths on round bar

Figure 18.35 shows how the vee block has been turned 90 degrees so a line can be scribed along the bar's length, which is parallel to the surface plate. This is used for marking bar to an exact length or to mark drilled holes or keyways.

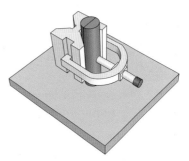

**Figure 18.35** Marking out lengths on round bar

## 66 Team Talk

Aisha: **'I have a gear wheel that is to fit over a shaft. How can I connect them both so that the gear can slide onto the shaft easily but then the gear and shaft both turn together?'**

Steve: **'You can use a key and key way. A small slot is machined into the shaft. The gear will already have a slot cut on its inside diameter. A long, thin bar, known as a key, is pushed into the shaft slot. This protrudes a couple of millimetres. The gear now slides over the key. This will mean that the gear and shaft will revolve together.'** 99

## Discs

Flat discs are more difficult to hold when marking out than cylinders or flat plate. They can be held in a vee block to ensure that the disc does not roll from side to side. An angle plate can be used to keep the disc vertical and perpendicular to the surface plate.

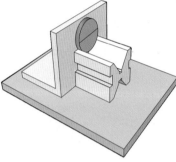

**Figure 18.36** Marking out discs

## 66 Team Talk

Aisha: **'Is it OK to lift an angle plate?'**

Steve: **'You need to be careful because they can be quite heavy. If you are unsure, do not lift any of this equipment and check with your supervisor.'** 99

### Make the grade

This activity will help you in achieving the following grading criteria:

**P4** mark out the three different applications to the prepared work plan;

**P5** demonstrate safe working practices and good housekeeping;

**M2** carry out checks to ensure that the marked-out components meet the requirements of the drawing or job description.

## Activity

To complete this activity, you need to select a product and create a drawing to be used for marking out. The potential number of products you could use is limitless; here is some guidance on the level of complexity required for marking out at this level. The drawings used for marking out should be of a relatively straightforward engineering product that may include the following features:

- parallel, angled or perpendicular lines;
- pitch circle diameters;
- circles and arcs;
- pockets and slots;
- simple dimensions (to one decimal place);
- chamfers;
- corners and radii.

The example shown in Figure 18.37 is simple but combines a range of typical features.

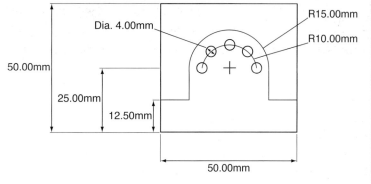

**Figure 18.37** Example marking-out drawing

Under normal circumstances, sizes are checked but not necessarily written down (or formally recorded). For the purposes of this activity, the checked dimensions could be recorded using Table 18.4.

When you have marked out a workpiece you should check the sizes you have marked out. Use table 18.4 to record your findings

| Marking out – check form | | | |
| Name of checker | | | |
| Dimension to be checked | Checked measurement | Accept or reject | Action to be taken |
| --- | --- | --- | --- |
|  |  |  |  |
|  |  |  |  |
|  |  |  |  |
|  |  |  |  |

**Table 18.4** Marking-out check form

# Marking out irregular-shaped workpieces

Marking out irregular- or odd-shaped components can cause a number of problems. These types of components can be:

- forged parts;
- sand-cast or die-cast parts;
- machined components.

Forged and cast components are not generally recognised as being produced to high levels of accuracy. These products may not have any obvious flat surfaces or datum to use as a starting point.

Figure 18.38 illustrates how to clamp and mark out a cast product. Castings and forgings can take various shapes and sizes, so it is tricky to provide a definitive guide.

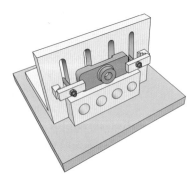

**Figure 18.38** Cast product

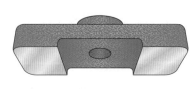

**Figure 18.39** Casting machined along base

In Figure 18.38, the front face of the workpiece is to be marked out. A parallel is used to lift the workpiece from the surface plate so it is easier to work with. An angle plate is used to ensure the workpiece is perpendicular to the surface table. The workpiece is an odd shape, so clamps with nuts and screwed bar (known as studding) are used to hold it in place. The face that is in contact with the angle plate is sand-cast and has a rough finish. To be completely accurate, this face could be machined flat using a machine tool such as a milling machine.

Figure 18.39 shows a casting that has been machined along its base. This will allow a more accurate setting.

# Learning Outcome 2. Be able to mark out engineering workpieces to specification

## Reading engineering drawings

Preparing for marking out is important. Before marking out commences, it is important to:

- fully understand the engineering drawing;
- identify a plan of marking out;
- select the correct tools;
- understand that the drawing will usually have dimensions in millimetres.

## Understanding engineering drawings

In order to complete any marking-out activity, it is essential to have a good understanding of engineering drawings. For starters, you need to know the various lines that commonly appear on an engineering drawing.

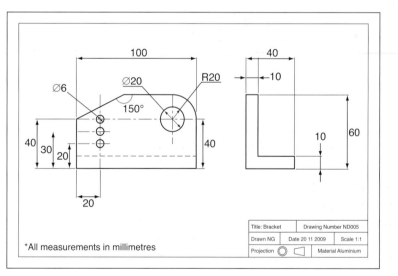

**Figure 18.40** Engineering drawing

## Outside line

This is a continuous line. It represents the outside shape of the component and is also used to represent holes and slots in the component. This type of line will be transferred onto the workpiece.

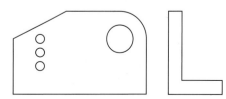

**Figure 18.41** Outside line

## Hidden detail lines

When a product is solid (and not transparent), you cannot see the details that are hidden from view. Engineers and designers represent this detail with a dashed line. The size of the line varies, but a typical 1:1 scale uses a 2 mm dash followed by a 2 mm gap, and so on. Figure 18.42 shows hidden detail lines to represent the flat material behind the component. On the left-hand side, the hidden detail lines represent the holes going through the material.

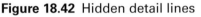
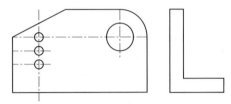

**Figure 18.42** Hidden detail lines

## Centre lines

These lines represent the centre position of holes or the centre of cylindrical features and components. On drawings, they are fine lines of long and short dashes with short gaps. When marking out on the workpiece, they should be light, but continuous, lines.

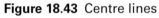

**Figure 18.43** Centre lines

## Dimension lines

On the engineering drawing, these lines represent the sizes, lengths, diameters, radii or angles of the product. Features are identifiable by the arrows at the end of each measurement. They are used to inform the engineer of the physical sizes to mark on the workpiece.

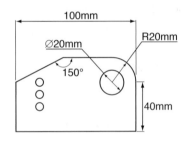

**Figure 18.44** Dimension lines

**66 Team Talk**

Aisha: **'How can you get light lines when marking out?'**
Steve: **'Make sure the marking blue only has a thin coating and then use a very sharp scriber. It can take a bit of practice to get lines thin.'** **99**

# Identification of types of material

This section will help you identify the main materials that are used in an engineering workshop. The engineering drawing will always state which material is to be used.

Materials to consider are:

- carbon steel;

- cast iron;
- aluminium;
- polymers (such as nylon or acrylic).

## Carbon steel

Steel is probably the most commonly used material in an engineering workshop.

**Figure 18.45** Carbon steel products

Carbon steel is a heavy material, dark silver in appearance, which is made by combining varying amounts of carbon with iron. Steel can corrode very easily due to the chemical reaction of oxygen (from the air) with iron (in the steel) to form iron oxide, or rust. As such, the steel will often be supplied with a thin coating of oil to prevent the onset of rust.

*Workshop products:* hand tools, clamps, drill drifts and vices.

## Cast iron

Cast iron often has a rough finish due to its high degree of hardness. It is supplied in billets and is not formed (and, therefore, not supplied) in flat sheet or round bar.

This is a heavy material with a rough finish. It is grey or dark grey in colour. It generally has a rough finish like sandpaper but it can also have very sharp edges created as part of the manufacturing process that produced the casting. It is hard and yet brittle, making it unsuitable to be formed in flat sheet.

*Products:* marking-out tables, brackets and vices.

## Aluminium

Silver in colour, aluminium is a light but strong metal. Aluminium does not go rusty, which is a great advantage.

*Products:* cans, foils, aircraft components and other lightweight components for bikes and cars.

## Polymers

There is a wide range of polymers (sometimes mistakenly referred to as plastics) available to engineers. Polymers are formed into products using highly automated processes such as extrusion. In an engineering workshop, where marking out is needed, a smaller range is generally used.

Polymers are lightweight, flexible and strong. The types usually found in the workshop include the following:

- **PVC** (polyvinyl chloride): often in tube or pipe form, this material is easily coloured.
  *Products:* cylindrical pipes and extruded products.
- **Acrylic:** usually supplied in flat-sheet form, acrylics can be transparent or easily coloured.
  *Products:* cases, boxes and business signs.
- **Nylon:** this is a hard and rigid material when it is in solid form. This means that it can be machined accurately and it keeps this accuracy, unlike some polymers that seem to bend easily. It is commonly supplied in round bar, in colours such as white, black and cream.
  *Products:* many gears, spacers, brushes and wheels in electrical components such as printers are made from nylon.

Polymers do not corrode, which is a great advantage, and their softness means that they can be marked out without the need for engineering blue. Of course, this also means that they can be easily damaged.

# Preparation for marking out

## Checking visual defects

Once you have decided on the type of material required, you will request this from the material stores.

Once you have received the material, the first job is to 'visually inspect' it.

Marking out is time consuming and therefore costly, so it is important to work effectively. One way to work effectively is to check the stock material before commencing the marking out to ensure it is free from imperfections or defects.

You should look for:

- cracks or splits in the material;
- dents;
- deformed materials, such as bent or compressed materials;
- small surface burns or melted polymers;
- corrosion;
- evidence that the material has previously been worked on;

- discoloured material, through heat treatment, which would affect the properties of the material.

This inspection is a quick process as any visual defects will be easily identified. If there are any defects, then return it to the store and ask for a new piece of material.

## Cleaning and preparation of the workpiece

Before marking out can take place, the material should be cleaned and prepared.

- Remove oil and grease.
- Remove protective coating.
- Remove burrs and flash.

### Remove oil and grease

Oil is often coated onto metal to prevent corrosion. Over time, this can become thick and dirty. This needs to be removed before marking out.

### Remove protective coating

When materials can easily be damaged, such as PVC or acrylic, or when the material has an aesthetic function, such as stainless steel, the material may be supplied with a thin polymer film covering the surface. This needs to be peeled off before marking out can take place.

### Remove burrs and flash

When materials are cut using power saws or guillotines, razor-sharp edges are created as well as rough spikes known as 'burrs'. Sharp burrs can easily cut a careless engineer. In addition, marking out would be inaccurate because the material may not sit perfectly flat on the surface plate. The burrs should be removed with a hand file.

Castings often have material that has seeped or overflowed during the casting process. This is known as 'flash' and has the same potential to cause harm and inaccuracy as burrs. Flash also needs to be removed with a hand file before marking out.

# Setting workpieces for marking out

## Setting workpieces

Once cleaning and checking of the stock material are finished, the process of marking out and setting can commence. The engineering drawing is the main 'tool' that is used to mark out. It must be referred to at all times.

## Creating a datum edge

When a piece of material is cut from a bar, this piece of material is known as the billet. This billet becomes the workpiece. On inspection, it is unlikely to be flat or square enough to start marking out. Therefore, you should:

- identify the datum edges;
- create scribed lines that will be filed;
- file the lines so the edges are flat and smooth;
- check for flatness and squareness.

Using the drawing, identify two edges from which all dimensions can be taken and which can be used as a reference.

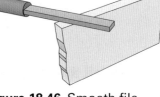

**Figure 18.46** Smooth file

The un-machined sawn edge of the material will usually be flat but not necessarily square and smooth. Use a smooth file to clean this edge. A smooth file with emery cloth covering the surface will give an even smoother finish.

> **Team Talk**
>
> Aisha: **'How can I get a really smooth finish on my workpiece?'**
> Steve: **'You should use emery cloth. It is a type of abrasive. It is a hard, gritty material glued to cloth. Make sure you file your workpiece as smooth as possible before you use emery cloth. It only removes tiny amounts of metal but gives a great finish.'**

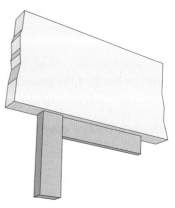

**Figure 18.47** Engineer's square

Use an engineer's square to check for flatness, along both the length and the width. The flat edge of the engineer's square should sit perfectly along the flat surface of the workpiece. If it is not flat, daylight will be seen through the gap.

When completely square, one datum edge has been created.

The workpiece can now be coated with marking blue. This will give the workpiece an even layer of ink, which can be removed with a scriber to create sharp, accurate lines.

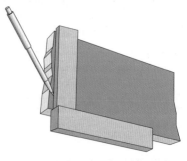

**Figure 18.48** Removing marking blue with a scriber

Using an engineer's square, a line can be scribed at exactly 90 degrees to the first datum edge.

Once filed and smoothed with emery cloth, the two datum edges form a perfect 90-degree angle. This can be checked with the engineer's square. If daylight can be seen through the gap, the protruding points that are touching the engineer's square will need to be filed off.

### Using slip gauges to help marking out

With workpieces that are odd shapes, it may not be possible to clamp the workpiece to an angle plate.

In Figure 18.50, the distance between the datum edge that rests upon the parallel and the front face of the boss are known (a boss is a profile that is raised up a little; in this case, it is the circle). Slip gauges are built up to the exact size to support the workpiece.

A dial test indicator is used to check that the top surface is parallel to the base. The combined use of a DTI and slip gauge facilitates the marking out of a range of irregular products.

# Marking out

D1

## Work plan

Always plan in what sequence you will carry out the marking out process by taking some time to consider the best or most appropriate way to mark out the workpiece.

We will use the drawing in Figure 18.58 as an example of effectively planning the marking out process.

### Marking-out operation 1

Mark all horizontal lines.

Equipment: vernier height gauge.

There are a number of horizontal lines. These are all produced at the same time, so it is best to mark them in the same operation.

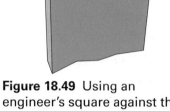

**Figure 18.49** Using an engineer's square against the first datum edge

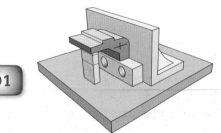

**Figure 18.50** Using a slip gauge to help marking out

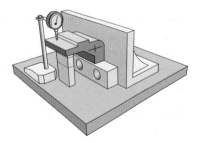

**Figure 18.51** Using a dial test indicator

## Marking-out operation 2

Mark 10 mm base all around workpiece.

Equipment: vernier height gauge.

Mark the 10 mm base all around the workpiece and up the sides of the workpiece as required. In this case, the size is consistently 10 mm, so all lines are marked in the same operation. This saves resetting the height gauge each time.

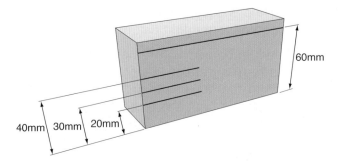

**Figure 18.52** Marking-out operation 1

This will give the L-shaped feature shown in Figure 18.54.

## Marking-out operation 3

Mark out 150-degree angle.

Equipment: vernier protractor.

The angle of 150 degrees is set on the vernier protractor. The left-hand datum edge is used. (Remember that the top edge in this case is not a datum.) Although the drawing stated 150 degrees from the top face, you must use the datum face and recalculate the angle appropriately. The angle from this face is 60 degrees.

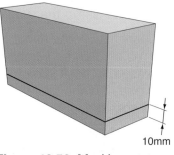

**Figure 18.53** Marking-out operation 2

## Marking-out operation 4

Mark out vertical dimensions.

Equipment: vernier height gauge.

The workpiece is now rotated 90 degrees so the vertical lines can be marked out. To save time, all the vertical lines should be done in the same operation.

**Figure 18.54** L-shaped feature

## Marking-out operation 5

Mark out radius and holes.

Equipment: dividers, centre punch and hammer.

In order to produce the radii and circles when marking out, you need to centre punch a small indent at the centre of the circle. This must be accurate and not too deep. It will allow the dividers to locate one point in order to scribe the circle.

The workpiece is fully marked out. Figure 18.58 shows the original drawing and Figure 18.59 shows the product machined out. Note how each line on the drawing matches the final product.

**Figure 18.55** Marking-out operation 3

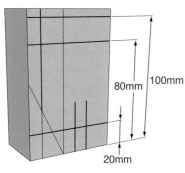

**Figure 18.56** Marking-out operation 4

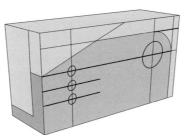

**Figure 18.57** Marking-out operation 5

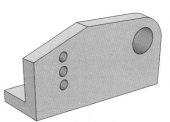

**Figure 18.59** Machined workpiece

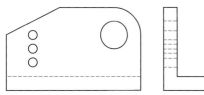

**Figure 18.58** Original marking-out drawing

| Marking-out operation | Description of operation | Equipment |
|---|---|---|
| | **Preparation** | |
| 1. | Clean material | Cloths |
| 2. | Check material sizes and de-burr | Steel rule, file |
| 3. | Mark out datums | Engineer's square |
| 4. | Ensure datums are flat and square | Engineer's square |
| 5. | Apply marking blue | Marking blue |
| | **Marking out** | |
| 1. | Mark all horizontal lines | Vernier height gauge |
| 2. | Mark 10 mm base all around workpiece | Vernier height gauge |
| 3. | Mark out 150-degree angle | Vernier protractor |
| 4. | Mark out vertical dimensions | Vernier height gauge |
| 5. | Mark out radius and holes | Dividers |

**Table 18.5** Work plan

## Make the grade

This activity will help you in achieving the following grading criterion:

**D1** justify the choices of datum, work-holding equipment and measurement techniques used to mark out the three different applications.

## Activity

Your teacher will supply you with drawings for three different products. The drawing below is a typical example.

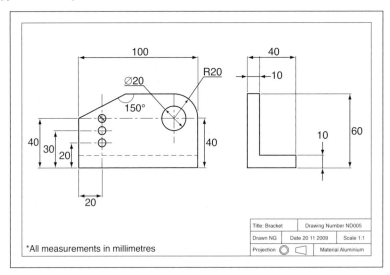

**Figure 18.60** Engineering drawing

Determine the most suitable datum for marking out the products. Different locations may be used as a datum, but it is your ability to provide justification for your chosen datum position that will be assessed.

## 66 Team Talk

Aisha: **'What is justification?'**

Steve: **'Justification means giving a reason why the datum position chosen is better than any alternative datum position.'**

Steve: **'To gain D1, it is important to justify techniques and marking-out equipment too. Again, this means a direct comparison with alternatives is needed.'** 99

# Safe working practices

There are lots of hazards in a workshop area. Personal protective equipment (PPE) must be worn for protection. This includes safety boots, overalls and goggles. There are some specialist areas, such as welding or heat treatment, where specialist PPE is used.

For all workshop activity you must have the following equipment:

- *Safety boots:* these are thick, leather boots with a steel toecap and thick, treaded sole. They help prevent slips, and protect against sharp objects around the workplace. The steel toecap protects against any object falling onto the toes.
- *Overalls:* these are an all-in-one suit that covers your everyday clothing. Overalls help prevent loose clothing and buttons becoming trapped in machines. They also protect clothes and skin from coolant and oils, which can cause irritation.
- *Goggles:* these are essential. They protect the eyes from sharp objects, airborne objects and splashing liquids. There are also many specialist types of goggles for welding and heat treatment. Always make sure you have the correct type and that your teacher or supervisor has seen them.

Marking out does not usually involve machinery, but there are a number of hazards that need to be considered, as follows:

- using tools with sharp points;
- using hammers, centre punches or letter stamps;
- creating datums using machine tools;
- metal filings, marking blue, and oils and greases;
- general marking-out environment.

## Using tools with sharp points

Scribers, dividers and odd-leg calipers have very sharp points and, if used inappropriately, can prove hazardous to the eyes or skin. Therefore, it is important to store them correctly. They should be laid carefully on the bench while being used and stored after use.

## Using hammers, centre punches or letter stamps

When using a hammer to hit centre punches and letter stamps there is a danger that the tools can slip out of your hand extremely quickly and can even fly across a workshop or up into your face. For this reason goggles should always be worn.

## Creating datums using machine tools

Sometimes datums can be formed with machine tools such as milling machines, centre lathes or grinders. It is not expected that you will create datums with a machine tool in this unit. Personal protective equipment (PPE) must always be worn when operating this equipment. This is covered in detail in Unit 14 (Selecting and using secondary machining techniques to remove material).

## Metal filings, marking blue, and oils and greases

Metal filings, marking blue, and oils and greases are all regularly found in the marking-out environment. There is a danger that they can contaminate the eye if the person marking out has them on their hands and accidentally wipes the eye. This would require at least basic first-aid treatment or, in some cases, hospital treatment. A further consideration is damage to the skin through regular use.

To avoid problems, you should keep your hands clean, and wash regularly, be careful not to spill or splash marking blue, and brush filings away from vices regularly.

## General marking-out environment

Generally, marking-out areas are clean and safe. However, you may be required to mark out in an industrial area. PPE may be required, so always be aware of safety signs.

## Barrier cream

Barrier cream is a product that protects the hands and can repair damaged skin. It should be applied every time you enter the workshop as it saves wearing gloves, which may hinder accurate processes such as marking out. Barrier cream prevents dirt, petroleum and marking blue from penetrating and damaging hands. It has anti-fungal and anti-bacterial properties that protect you while you work. Regular use helps repair damaged skin. If you have a reaction to a barrier cream or are unsure if you should use it due to an allergy or skin condition such as dermatitis, you should ask your supervisor.

# Workshop behaviour

## Workshop behaviour

Workshops are dangerous places and accidents can happen at any time. Keeping you and the other workers safe is not only important, it is the law!

Following some simple rules will help keep everyone safe.

- Never run.
- Use designated walkways in the workshop.
- Never fool around.
- Only use marking-out and other tools for the purpose for which they were designed.
- Never sit on workbenches.
- Do not put sharp tools in your pocket.
- Do not touch machinery unless you are properly trained and authorised.

## Health and safety risks to others

What you do in a workshop can affect the safety of others in the workshop. You need to consider the following:

- Return tools to their designated storage place.
- Clean up your work area after use.
- Behave safely at all times.
- Do not throw materials or tools.
- Report any problems that you think are unsafe.
- Do not block walkways with objects.
- Report any accidents or near misses.

**66 Team Talk**

Aisha: **'Why should we report a "near miss" if nothing happened?'**
Steve: **'Nothing happened this time but next time we might not be so lucky!'**

**Make the grade**

The next activity will help you in achieving the following grading criterion:

**P5** demonstrate safe working practices and good housekeeping.

## Activity

Ask your teacher to complete the checklist in Table 18.6 to confirm that you are working safely and are carrying out good housekeeping. This could also be applied a number of times to show consistency. Your teacher should describe your performance and give evidence of outstanding practice or recommendations for improvement.

Photographic evidence could be used to support this checklist.

| Marking out – safety and housekeeping | | |
|---|---|---|
| The student was observed and was able to: | Yes/No | Action required |
| ensure work area was clean before use | | |
| lay out tools carefully | | |
| inspect tools for fitness for purpose before use | | |
| use tools safely | | |
| behave safely at all times | | |
| clean and store tools after use | | |
| clean up work area after use | | |
| Description of student's performance and recommendations | | |

**Table 18.6** Marking out – safety and housekeeping

# Housekeeping

Housekeeping means making sure that work areas, machine tools, hand tools and equipment are all kept in a safe and organised manner.

Disorganised areas cause a number of problems. You need to consider:

- safeguarding tools;
- avoiding slips and trips;
- keeping workpieces safe.

In workshops, there are guidelines that will help prevent these problems.

## Safeguarding tools

Tools should be cleaned and returned to their designated place after use. Usually, tools and equipment will have their own box or will be placed in a cabinet or drawer after use.

### Storage of devices

All marking-out tools and devices have their own designated storage places. This keeps the devices clean, safe, in good working order and accessible for the next user.

| Scribers and centre punches | These can be kept in a wooden block into which holes are drilled. This keeps the points safe and clean while providing easy access. |
|---|---|
| Engineer's squares | These are usually stored in individual boxes but sometimes in a flat drawer, which provides quick and easy access. |
| Dial test indicators | These are usually kept in individual cases because they are very sensitive. |
| Vernier height gauges | These are usually stored upright in a cupboard. |
| Marking-out tables/surface tables | These should always have a protective cover to prevent damage. The surfaces should be wiped clean after use. |

**Table 18.7** Typical methods of storage

# Avoiding slips and trips

At the end of each workshop session, the area should be cleaned. This will involve using a hand brush to clean down the benches and remove metal filings and off-cuts of material. Sweeping the floor will ensure that the work areas and walkways are in good condition and free from slipping hazards.

## Disposal of waste

Common waste that is left in a workshop area can cause a number of problems.

- *Material off-cuts:* these are sharp and can easily graze, pierce or cut the skin. They should be swept up and placed in the bins provided. If there are separate bins for different materials, be aware – metals and polymers are often recycled.

- *Cleaning cloths:* these are used to wipe oil and dirt from materials and should be thrown away in the bins provided. The oil and dirt in these cloths contain bacteria, which may cause skin irritations if not disposed of properly.

- *Other materials:* hacksaw blades, emery cloths and empty marking-blue bottles may all be found at the end of the session – all need to be disposed of responsibly. Hand towels (used for drying hands after a workshop session) can cause a slipping hazard and must be disposed of correctly in the bins provided.

# Keeping workpieces safe

Workpieces should be kept safe at all times. At the end of any workshop session, your name should be clearly marked on your workpiece. This could be hand stamped, scribed or simply written on with a marking pen. Where the job involves more than one part, all parts should be put in a clearly marked bag or taped together with masking tape.

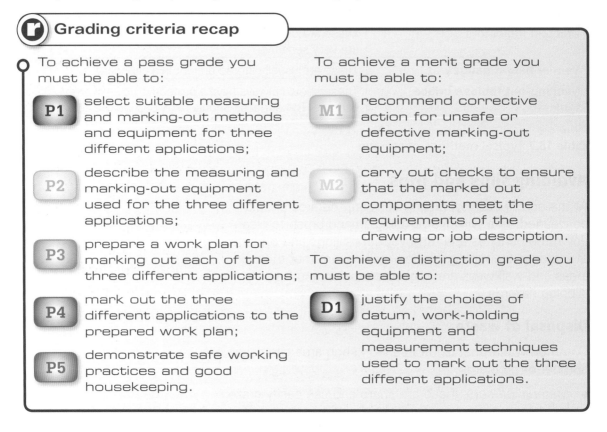

**r** **Grading criteria recap**

To achieve a pass grade you must be able to:

**P1** select suitable measuring and marking-out methods and equipment for three different applications;

**P2** describe the measuring and marking-out equipment used for the three different applications;

**P3** prepare a work plan for marking out each of the three different applications;

**P4** mark out the three different applications to the prepared work plan;

**P5** demonstrate safe working practices and good housekeeping.

To achieve a merit grade you must be able to:

**M1** recommend corrective action for unsafe or defective marking-out equipment;

**M2** carry out checks to ensure that the marked out components meet the requirements of the drawing or job description.

To achieve a distinction grade you must be able to:

**D1** justify the choices of datum, work-holding equipment and measurement techniques used to mark out the three different applications.

# Unit 19
## Electronic circuit construction

# Introduction to the unit

Over the past few decades, the simple electronic circuit has rapidly become a vital part of our everyday lives. Whether we are socialising with friends or carrying out a simple task at work, an electronic device will never be far away.

This unit will introduce you to the skills and knowledge needed to work as an electronics technician. To do this, we will start from scratch, explaining the operation of some basic components and the symbols used to represent them. The next step will be to produce simple circuit diagrams before examining the techniques used to transfer our diagrams to a stage of production. A series of simple circuits will then be explained and, finally, some circuit construction techniques will be discussed. At various points throughout the unit, we will examine the safe practices needed to work with electronics.

## Learning Outcomes

By the end of this unit you should:

- be able to use safe working practices in the electronic laboratory/workshop;
- know about electronic components and circuit diagrams;
- know about the manufacture of electronic circuit boards;
- be able to construct an electronic circuit.

# Grading criteria

| To achieve a pass grade you must be able to: | To achieve a merit grade you must be able to: | To achieve a distinction grade you must be able to: |
| --- | --- | --- |
| **P1** describe the potential hazards related to constructing electronic circuits | **M1** explain the function and operation of four different electronic components | **D1** propose a method used to construct a given electronic circuit and justify your choice |
| **P2** use safe working practices in the electronics workshop/ laboratory | **M2** explain the advantages and disadvantages of the three types of electronic circuit board | |
| **P3** describe the purpose of six different types of electronic component | | |
| **P4** read a given circuit diagram to identify the electronic components in the circuit | | |
| **P5** describe the manufacture of the three types of electronic circuit boards | | |
| **P6** use two methods of construction for a given electronic circuit | | |

# Learning Outcome 1.  Be able to use safe working practices in the electronic laboratory/workshop

## Hazards and safe working practices in the electronics laboratory

As with many engineering operations, electronics has its fair share of associated hazards. What often makes these hazards more dangerous is that they are hidden; for example, it is impossible to tell if a piece of wire has electricity flowing through it just by looking at it. Therefore, when working with electronics, it is vital to understand how it should work, the hazards that may occur and, most importantly, how to reduce the hazards. Many risks can be established using a procedure called risk assessment. This procedure is discussed in Unit 1 (Working safely and effectively in engineering).

### Cuts and abrasions

The most common cause of cuts and abrasions when working with electronics is hand tools. As an electrician or an electronics technician, your tools will be one of your most important assets. They can play a major part in the quality of your work and how quickly you complete it. To reduce the risk and ensure the safe use of hand tools, the advice can be summed up in three main points.

1. Always check the condition of hand tools before use and, if damaged, they should be replaced. This sounds pretty obvious but do not adopt the 'It will do' attitude. Tools can easily become damaged, such as blunt pliers, damaged test leads and snapped screwdriver tips. All of these are potential hazards.
2. Always use hand tools for their intended purpose. If there is a task to do, the chances are a hand tool has been invented to make it easier. Using the wrong tool will only make the task take longer, cause damage or create a hazard.
3. After carrying out a task, make sure your hand tools are tidied away safely and stored in a clean and organised manner. This will ensure that you can quickly get the tool you need next time and that it will be in a safe and useable condition. In effect, this point will ensure that points one and two above are met.

It is often necessary to trim the legs of components once they are secured onto a circuit board. In order to reduce the risk of scratches from these exposed legs, a pair of side cutters should be used to trim back the legs as close to the circuit board as possible. Any remaining strands of wire should then be cleaned from the work area.

While constructing an electronic circuit, it may be necessary to use a small-diameter drill bit to create mounting holes in a circuit board. Due to the small size of the drill, very fine swarf is produced, which should be swept away and removed from the drill bit after all holes have been drilled. This prevents the swarf causing a short circuit. More information on how to identify and reduce the risk of hazards can be found in Unit 1 Working practices.

## 66 Team Talk

Aisha: **'Is there another way I could store my tools rather than in a tool box?'**

Steve: **'You could create a tool wall or shadow board. Good DIY stores sell mounting hooks, which you can use to store your tools on a garage wall, and some tools, such as adjustable spanners, even have mounting holes. Storing your tools this way also allows you to see if there are any missing because there would be an empty space on the wall.'** 99

## Soldering hazards

Soldering irons are arguably the most common cause of burns while working with electronics. The tip of the soldering iron once fully heated can reach a temperature of 400°C. This would give you a nasty burn if it came into contact with your skin. To reduce this risk, the soldering iron should be kept in a soldering stand when not in use; it should never just be laid on a table. Remember that the soldering iron can take up to ten minutes to cool down after being unplugged.

You will notice that, while you are soldering, a smoke is produced; these fumes are caused by the melting flux inside the solder and they can be harmful to the eyes over a prolonged period. Inhalation of these fumes is not advisable as solder contains lead. To reduce this risk, always solder in a well-ventilated area and wear eye protection. If you are soldering for long periods of time, extraction systems are essential to remove the hazardous fumes.

## Risk of electric shock

Arguably the most common risk associated with working on electronic circuits is the possibility of receiving an electric shock. All electrical components need some form of electrical supply to work; whether it comes from a battery or a mains socket, a risk is always present.

An electric shock occurs when an electrical current passes through your body. The term 'current' is used to describe the flow of electric charge. A shock happens when a part of the human body comes into contact with a live piece of electrical circuitry. The main point to stress is that it is the current that causes the damage, not the voltage; remember the phrase 'Current kills'. The term 'volts jolts' is used to remind us that it is a high voltage that makes us jolt when we touch it, but it is a high current that presents a danger. But what damage does current cause?

There are three main ways that electricity can hurt you:

1. Neurological – electricity interferes with our nervous control system and can cause anything from a slight tingle to a heart attack.
2. Burns – low-voltage electrical burns can be caused on the skin, but high-voltage burns can be inflicted to internal organs.
3. Ventricular fibrillation – this is a cause of cardiac arrest, meaning it makes the heart stop pumping blood. Ventricular fibrillation is caused by a current flowing through the chest and is basically the procedure used to bring a cardiac-arrest patient back to life. (A defibrillator stops the heart from pumping blood so that chest compressions can begin safely.)

By far the simplest and most effective way to reduce the risk of electric shock while working on a circuit is to switch off the circuit and then fully disconnect the electrical supply. Not all circuits have an on/off switch and, therefore, it is vital to remove all forms of power supply. An electrical meter can then be used to check if a circuit is fully 'dead' or 'isolated'.

## Etching-fluid hazard

Ferric chloride is a common chemical used during the etching process to produce copper printed circuit boards. Using this chemical presents four main hazards:

1 Contact with the skin – etching fluid is highly corrosive and it will immediately burn the skin or eyes on contact. To prevent this, gloves, goggles and protective clothing should be worn at all times when working with etching fluid.

2 Gas inhalation – the process of etching produces a toxic gas, which is very harmful if inhaled. To prevent this, good ventilation is required; if etching is being carried out on a larger scale, a full extraction unit must be in place.

3 Swallowing – the effects of swallowing etching fluid will depend on the amount consumed. To control this hazard, etching tanks should be covered at all times and replacement chemicals stored in a secure location.

4 Disposal – to meet the COSHH regulations for the used fluid, it is often the case that the fluid must be collected and disposed of by specialist disposal contractors. This can have financial implications for some companies.

## 66 Team Talk

Aisha: **'When I give someone an electric shock by walking across a carpet and then touching them, how come it doesn't cause any real damage?'**

Steve: **'This is caused by static electricity and the shock voltage can be as high as 10,000 volts, which is why it tingles. The current is so low and only present for milliseconds so no damage is caused.'** 99

## Activity

Using the internet, research the current levels that are applied to the body during an electric shock and identify the effects they cause to the human body. Record your findings in a table (Table 19.1). This activity can be carried out in small groups and the results discussed at the end of the activity.

| Current value | Effect on the human body |
|---|---|
| Less than 1 mA | |
| Between 3 and 4 mA | |
| Between 5 and 10 mA | |
| Between 30 and 50 mA | |
| Over 50 mA | |

**Table 19.1** Current levels and effects on the human body

## Use of first-aid procedures

If an accident does occur in the electronics laboratory, a company-certified first-aid representative should be called to ensure that the most appropriate course of medical care is taken. However, there are a few simple first-aid procedures that can be followed to treat the following accidents:

1. Acid burns – immediately wash the contact area with cold water and remove any contaminated clothing. If eye contact is made, rinse immediately with cold water or eye wash and seek medical attention. If swallowed, seek immediate medical attention.
2. Electric burns – immediately cool the burnt area with cool running water for at least ten minutes. Then remove any jewellery and cover the area with a clean sterile dressing.
3. Electric shocks – immediately switch off the electricity supply or break the electrical contact between the conductor and the patient using an insulating material such as wood. You should now call for assistance and place the patient in the recovery position.

## Personal protective equipment (PPE)

As with many engineering operations, the hazards caused when working with electronics can be reduced by wearing appropriate personal protective equipment (PPE). Some guidelines for the use of PPE are explained below.

- Safety glasses should be worn when carrying out any drilling operations; this is regardless of the size of the drill bit. They can also be worn when assembling electronic circuits as clipped wires often fire in random directions; they also keep soldering fumes out of the eye, which can often cause irritation. Safety glasses will also protect the eyes when soldering or desoldering. Wires or leads can often spring apart, splattering molten solder into the eyes or face.
- A lab coat can be worn to remove any dangers caused by loose-fitting clothing; it will also act as an extra barrier to protect the skin from any burns, chemicals or oils.
- Metallic jewellery should also be removed to prevent the risk of it causing a short circuit or touching the soldering iron.

## Cable colour coding

Many guidelines and regulations cover electrical equipment, all of which are designed to protect us, as users, and also technicians carrying out repairs. One good example of this is the cable colour-coding system. In January 2005, Europe standardised the colour coding system for cables used in domestic wiring. This standardisation meant that any person working on domestic wiring would always be able to correctly identify the line or phase, neutral and earth conductors, regardless of the size and type of cable used. This would prevent any confusion. Previously, flexible cable had a different colour system. Table 19.2 shows the new and old colour-coding system.

| Conductor | Pre-2005 colour | New standard colour |
|---|---|---|
| Line or phase L | Red | Brown |
| Neutral N | Black | Blue |
| Earth I | Green | Green/Yellow |

**Table19.2** Cable harmonisation

Figure 19.1 shows a three-core mains cable with the PVC insulation stripped back to reveal the solid-copper conductors.

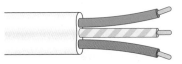

**Figure 19.1** Three-core mains cable, stripped back to reveal the line or phase (brown), earth (green/yellow) and neutral (blue) copper conductors

## Fuses

The fuse is a very simple device that carries a great deal of responsibility. The role of the fuse is to protect both the electrical equipment and ourselves in the event of a fault. It is more common to use a trip device such as an over-current protection device, but a fuse is the cheapest and simplest solution.

The construction of a typical fuse is simply a thin piece of wire secured in some sort of casing. This special wire is designed to melt at a set temperature. Under normal conditions, the fuse can get rid of or 'dissipate' the heat generated by an electrical current, but when a fault occurs in the electrical circuit, the current going through the wire starts to increase. Eventually, the heat going through the wire gets so high that the wire melts; this is known as 'blowing the fuse'. This 'blowing' action causes a break in the supply, therefore turning off the supply to the electrical circuit.

Fuses come in many shapes and sizes, depending on the circuit they are protecting. A domestic BS1362 plug fuse is shown in Figure 19.2.

The value of these fuses can be 3, 5, and 13 amps, which is their current rating. When this rating is exceeded by 40 or 50 per cent, they will blow. Plugs usually come with a 13-amp fuse fitted as standard; this must be replaced to match the current rating of the cable being protected.

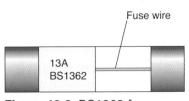

**Figure 19.2** BS1362 fuse

## Activity

Remember, fuses are rated in amps (A). This is the unit used to measure the flow of current. We can calculate the value of fuse needed by looking at the wattage and voltage of the appliance being tested and applying the simple formula below.

$$\text{Amps} = \frac{\text{watts}}{\text{volts}}$$

We can then choose a fuse that is closest to the calculated amperage; never use a fuse that is too high as this may cause a fault to start a fire.

Table 19.3 contains some typical appliances found in the laboratory. Using their supply voltage and power ratings in watts, calculate the fuse value required.

| Appliance | Voltage (V) | Watts (W) | Calculated amps | Suggested fuse (A) |
|---|---|---|---|---|
| Soldering iron | 230 | | | |
| Table lamp | 230 | | | |
| Computer | 230 | | | |
| Electric heater | 230 | | | |

**Table 19.3** Typical appliances and fuse values

Now that the fuse values have been calculated, can you suggest a general rule for selecting a fuse based on the following appliance ratings?

- Appliances up to 700 watts: use ____ amp fuse.
- Appliances from 700 to 1,200 watts: use ____ amp fuse.
- Appliances from 1,200 to 3,000 watts: use ____ amp fuse.

## Replacement of a mains plug

Portable electrical appliances are fitted with a three-pin plug, like that shown in Figure 19.3.

The plug is used to provide the electrical connection between the appliance and the mains supply, and it also houses the fuse. The three pins on the plug are the line or phase, neutral and earth. On the plug in Figure 19.3, the pins are made from brass, which is a good conductor of electricity. Looking face-on at the pins of the plug, the top pin is the earth, the left pin is the line or phase and the right pin is the neutral. You can see that the earth pin is longer than the other two.

**Figure 19.3** Three-pin plug

There are two reasons for this, both concerned with safety. Being the longest pin, the earth will be the first pin to make a connection and the last to break connection; this ensures the appliance is earthed before any voltage is applied. Secondly, the earth pin is needed to open the socket shutters. These shutters block the live and neutral connections and stop small objects being placed into the sockets.

**Figure 19.4** Plug internals

It is common these days for appliances to be supplied with moulded plugs, but accidents can happen and plug casings or pins can become damaged. When this happens, the old plug can be easily removed and a new plug fitted using the simple steps below.

1.  Remove the old plug. If it is a moulded type, the cable will need to be cut as close to the plug as possible to keep maximum cord length. If it is the changeable type, first remove the top cover by unscrewing the centre screw, using a medium-sized screwdriver. The screw need not come all the way out but, if it does, remember to put it somewhere safe. Figure 19.4 shows how the internal connections of the new plug should look.
2.  Some plugs also have a cord grip, like that shown in Figure 19.4. This is used to secure the cable and prevent any stress being placed on the terminals in the event of the cable being pulled. Completely remove one screw, using a terminal screwdriver, and slacken the other screw so that the grip can be rotated out of the way.
3.  Now use the terminal screwdriver to slacken the three brass terminal screws and completely remove the old plug.
4.  It is good practice to make new connections, so cut off the cable level with the outer insulation. Lay the cable on top of the plug with the end level to the back of the earth pin terminal; now place your thumb just above where the cord grip will be to mark where the outer PVC sheath needs to be stripped to. Use a pair of wire strippers to carefully remove the outer sheath. Take care not to damage the wire insulation.
5.  You now need to cut each conductor to length. First check the earth length is OK, because this is the longest conductor. All the conductors should lie nicely in the plug without being stretched. Now lay the blue neutral connector in place and cut the cable level with the back of the terminal. Then do the same with the brown line conductor.

6. Using the wire strippers, remove about 8 mm of insulation from each conductor and twist the stranded copper wire using your fingers. Now double the wire back so that the stranded copper is folded neatly in half.
7. At this stage, it is best to secure the cord grip. Put the cord in place and swing back the grip. Ensure the grip lies over the cord, as shown in Figure 19.4. Insert the removed screw and tighten each in turn so that the grip lies parallel.
8. Now place all wires in their appropriate positions, ensuring the wires lie neatly in the plug. Secure the terminals, making sure all copper strands are neatly in the holes and the conductor insulation is level with the front of the pin terminal, as shown in Figure 19.4.
9. The plug cover can now be replaced and secured with the screw.

## Activity

Replacing or fitting a 230V three-pin plug is a common task for an electronics engineer. Using the procedure listed in the text, fit a three-pin, 230V plug to a length of a 3 core pvc/pvc flex. Take a photograph at each stage of the process and produce a poster to demonstrate the correct safe working procedure.

## 66 Team Talk

Aisha: **'I have seen plugs before with round clamp terminals. Are they changed the same way?'**
Steve: **'Yes, just make sure that that the copper conductor is neatly twisted round the post in a clockwise direction. That way it will not come loose when you tighten the screw.'** 99

## Checking earth connections

The earth connection or circuit protective conductor (CPC) helps protect the circuit and the user in the event of a fault. Therefore, before any form of electrical supply is applied to a circuit, it is good practice to check the earth connections. The most efficient way to check this is to carry out a continuity test using a suitable multimeter in resistance mode. A continuity test is used to prove that the CPC has a low resistance and is continuous throughout the circuit. This will ensure that, in the event of a fault, the large fault current will flow through the CPC and operate the fuse of the circuit breaker, isolating the circuit.

## Checking polarity

When working with electronics, it is important to understand the concept of polarity. Electrical devices such as power supplies, cells, batteries, polarised capacitors and semi-conductors have polarity and this determines how they are connected into a circuit. Electrical polarity describes the path of electron flow through a circuit, and the electrons carry either a positive (+) or negative (−) charge. The electron flow is always from the negative to positive pole. In a typical electronic circuit, direct current (DC) will be used to supply the electron flow in one direction only. It is therefore vital that all of the electrical devices used within the circuit are connected with the + and − terminals in the correct positions.

**Figure 19.5** Electrical tools

## Portable appliance testing

Commonly known as PAT testing, this is a process where electrical appliances are routinely checked to confirm their electrical safety. If an appliance passes the test a 'passed' label is clearly fixed to the appliance. 'Do not use' labels will also be visible if the appliance has failed the test.

## Hand tools

Now let's look at some of the typical tools you are likely to use and illustrate why you would use them. The tools described can be seen in Figure 19.5.

### Screwdrivers

The most common screwdriver used by an electrical technician is the insulated type. This allows you to hold the blade without the risk of electric shock. Screwdrivers come in various sizes, from terminal through to large, and with flat heads or pozidriv®. The best option is to select a good range of types to allow the most suitable screwdriver to be selected based on the screw head.

### Side cutters

Side cutters or snips are a very important tool. Snips are used for cutting small cables and single conductors or wire. They can often be used to strip insulation, although care must be taken when doing this to prevent the possibility of cutting too deeply or notching the conductor, creating a weak spot. This could cause the conductor to snap when it is secured. For electronics, a small, insulated pair of snips would be most suitable.

## Wire strippers

Wire strippers can be expensive, but they can save a great deal of time and remove the risk of damage to conductors. They are also suited for use on very small conductors, which can be tricky to strip.

## Pliers

Pliers have many uses and come in many forms. In electronics, the most common are 'long-nosed' (sometimes called 'snipe-nosed') and 'flat-nosed' types. Both can be used for gripping or cutting cables, bending conductors and holding components. Pliers are not intended to be used to strip wire. As with all electrical tools, pliers with insulated handles are most suitable.

## Soldering iron

Soldering is a joining process used in electronics to secure electrical connections. The process involves using a hot soldering iron to heat the two conductors that need joining and then fusing them together using a soft solder, which melts at about 200 degrees Celsius. As the solder (an alloy of tin and lead) cools and hardens, the two pieces are joined together. Soldering irons come in various sizes and can be powered by gas or electricity. A small soldering electric iron with a power of around 30 watts is most suitable. Soldering iron stands can also be purchased to safely store your iron while in use.

In order to minimise the risks associated with using a soldering iron, the following steps can be used as a guide when carrying out simple soldering exercises.

1. Selecting a good soldering iron is the first step. Inspect the tip for damage and make sure it is a suitable size. If you are soldering small components to a circuit board, then a small-tipped iron will be the easiest to use. Once you have made your choice, place the iron in the stand and turn it on. While the iron is heating up, soak the small cleaning sponge in some clean cold water.

2. Inspect all joints to be soldered for grease and dirt. These can act as a barrier when soldering and reduce the quality of the join. Any cleaning that needs to be done can be carried out using wire wool, fine emery cloth or a circuit-board eraser. You can also clean the

tip of the soldering iron by removing it from the stand and pressing it firmly onto the sponge. Do not hold it on too long or the sponge will burn.

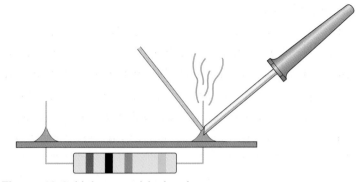

**Figure 19.6** Using a soldering iron

3. The components to be soldered can now be positioned in the circuit board. If it is possible, use a clamp or work-holding device such as a jig to secure the board steady because this will make the task easier. Add a small amount of solder to the tip of the iron so that it forms a neat pool of solder across the tip of the iron. If the solder only adheres to certain parts of the tip, then this indicates that it is too dirty and may need cleaning or replacing.

4. Touch the tip of the soldering iron against the copper track and the component leg. After a few seconds, feed a small amount of solder into the component and track so that it flows around the join, forming a cone shape, as shown in Figure 19.6. Once done, remove the solder first, then the iron and allow the joint to cool before disturbing any components or connections.

5. Once all of the circuit soldering is complete, clean the tip of the iron and return it to the stand before turning it off. A pair of wire cutters can then be used to trim back any wires or component legs protruding from the circuit board.

## 66 Team Talk

**Aisha: 'What is the difference between a Phillips® head screwdriver and a pozidriv® screwdriver?'**
**Steve: 'The easiest way to tell the difference is to look at the tip of the screwdriver. On the Phillips®, it simply looks like a cross; on the pozidriv®, the tip has a cross and additional lines at 45 degrees. Try looking up an image on Google.'**

## Make the grade

The aim of this Make the grade is to allow you to demonstrate that you are fully aware of the hazards relevant to the construction of electronic circuits. The hazards explored will be associated with the soldering of components. The next activity will help you in achieving the following grading criterion:

**P1** describe the potential hazards related to constructing electronic circuits;
Unit 4 (Applied electrical and mechanical science for engineering) provides further information on current, voltage and power.

## Activity

The joining process of soldering may be simple but there are many hazards and risks that you should be aware of. Your task is to produce a visual A4 poster, which can be displayed around the lab, to highlight the correct soldering technique and the safety procedures that should be followed. In groups, use the library or internet to obtain all the relevant information and look at the design of similar posters.

# Learning Outcome 2. Know about electronic components and circuit diagrams

## Electronic components

Before we start to look at individual electronic components, we need to recap a few basic terms. An electronic circuit is simply a collection of components attached together to carry out a desired operation. There are many electronic components available and a selection of those most commonly used will be described in this section.

First of all, let's consider the materials used to make electronic components. They fall into three categories:

1. Conductors – these materials allow the flow of electricity, e.g. copper.
2. Insulators – these materials block the flow of electricity and usually cover the conductors to provide mechanical protection, e.g. polymers such as PVC.
3. Semi-conductors – these materials are a mixture of the above and have conducting and insulating properties depending on how they are connected into the circuit. Silicon is the most common semi-conducting material; to create the semi-conductor properties, other elements are added in a process called 'doping'. Transistors, diodes and integrated circuits (ICs) are all made from silicon.

## Cell

A simple cell is a source of electrical power called direct current (DC). The electricity is produced because of a chemical reaction taking place inside the cell. Once this reaction is complete, the cell is either dead and cannot be recharged (primary) or restored if it is the rechargeable type (secondary). The most common cells produce 1.5V and are used to power clocks, torches, calculators, digital watches and electrical toys. A battery is the name given to a series of cells connected together.

## Battery

A battery is made by connecting a number of cells in series to make the voltage needed. The common 9-volt square battery is made by joining six 1.5V cells. This can be seen in Figure 19.7.

## Fuse

A fuse is a protection device fitted in the line voltage supply wiring or circuitry to protect both the electrical equipment and ourselves in the event of a fault.

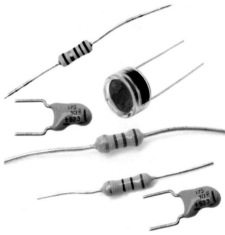

**Figure 19.7** Battery and cell

The fuse contains a thin piece of wire secured in some sort of casing. All current flow produces heat and a fuse is selected so that, under normal conditions, it can dissipate this heat. However, if a fault occurs, the current going through the wire starts to increase. Eventually the heat going through the wire gets so high that the wire melts and the supply is broken, therefore protecting the circuit components from damage.

## Resistors

It is obvious from the name that resistors are used to resist something. In an electrical circuit, we can use a resistor to restrict the flow of electrons. The ohm is the unit used to measure this resistance, represented by the symbol $\Omega$.

Resistors can be divided into three main types, which are outlined below. Some typical resistors are shown in Figure 19.8, such as LDR, fixed and variable resistors.

### Fixed resistor

**Figure 19.8** Resistors

This type of resistor offers a fixed value of resistance that cannot be altered; this is the most common type of resistor and it can be found in practically every electronic circuit. The fixed resistor can be used in a circuit for two main purposes. It can protect another component by restricting, limiting or reducing the current flow to a safe level. An example of this is to protect an LED. It can also split voltages within a circuit by acting as what is called a voltage or potential divider. In this type of circuit, the resistor causes the voltage to be split to two different values. This means that if a 9-volt battery is used, a voltage divider can create a 6-volt supply without the need for another battery.

## Variable resistor

As the name suggests, this type of resistor has the ability to vary its resistance. The most common example of this is the volume control of a car stereo, which needs to be continuously altered, or a light dimmer switch. They can also be used in voltage-dividing circuits, allowing the voltage to be fine-tuned. Variable resistors are made using a fixed resistor element – a sliding contact is then added, which moves up and down this element, giving the ability to vary the resistance needed.

## Thermistor – special function resistor

One commonly used resistor with a special function is a thermistor. All resistors are affected by a change in temperature and a thermistor uses this to its advantage. The thermistor is specially designed so that, when the surrounding temperature changes in value, its resistance changes in value. These properties allow the thermistor to be commonly used in electronic circuits that need a temperature sensor. Some examples of this are a central-heating control circuit that uses a thermistor to monitor room temperature, or the temperature monitoring of an electric motor.

## Light-dependent resistor (LDR) – special-function resistor

As the name suggests, this resistor has a special function related to the sensing of light, meaning the resistance value of an LDR will change due to the changes in light conditions of its surroundings. Most LDRs are designed so that, as the light increases, its resistance value decreases. LDRs are commonly used as sensors in electronic circuits that switch on lights automatically when it becomes dark.

# Diode

The diode is a simple, semi-conducting electronic component with a specific function; it allows current to flow through it in only one direction. It can be used in an electronic circuit in two ways, depending on how it is connected. In forward bias mode, it will allow current flow; in reverse bias mode, it will block current flow. The diode is commonly used in what we call rectifying circuits; these circuits are used to convert an AC voltage into a DC voltage. Some typical diodes such as an LED, Zenner and bridge rectifier are shown in Figure 19.9.

### Light-emitting diode (LED)

The LED is a special type of diode that also allows current to flow through it in only one direction. However, it has a difference. It is specially designed so that when it is connected in forward bias mode, thereby allowing current flow, a light is generated (emitted). LEDs are available in various colours and are gradually replacing the common bulb, reducing cost and increasing efficiency.

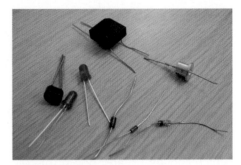

**Figure 19.9** Diodes

## Transistors

The transistor basically operates as a simple switch, meaning that it switches on and off as required. The transistor is available in many forms but the most common is the bipolar junction transistor (BJT). This type of transistor is made from silicon and has three terminal connections. There are two types of transistor: NPN and PNP. These have different symbols (as shown in Figure 19.11). The letters refer to the type of material used to make the transistor. Both types have leads to the base, collector and emitter.

**Figure 19.10** Transistors and capacitors

In the NPN type the base connection is used to switch the transistor on and off; when a small current is applied to the base, the transistor turns on. When turned on, the transistor allows a larger current to flow from the collector to the emitter. The transistor is commonly used in amplifier circuits, as a small base current allows a large collector–emitter current. The transistor gain is the name given to describe the ratio of output to input for these currents.

Another type of transistor commonly used is the field effect transistor (FET). The FET operates in a similar way to the BJT but it has several properties that make it distinctive. For example, a FET does not need a protective resistor because it has built-in protection. It also operates over a range of applied voltages, typically 3V to 15V, and the base, collector and emitter connections are replaced by gate, drain and source. A common use for FETs is switching analogue signals.

Figure 19.10 shows some typical BJT transistors and capacitors used in electronic circuits.

## Capacitors

A capacitor is an electronic device used to store an electrical charge. Like resistors, they can be fixed or variable, and the bigger the capacitor, the more electrical energy it can store. The capacitor is made using a sandwich of two plates with an insulating layer in between, called a dielectric. The bigger the plates, the bigger the charge that can be stored. When electricity is supplied to a circuit containing a capacitor, it begins to charge; when the supply is removed, the capacitor discharges. This can often be seen when switching off an electrical device and a coloured LED slowly fades. Capacitors are used in timing circuits and also for smoothing rectified AC voltages.

The unit used to represent a capacitor's value is the farad. In typical electronic circuits, only small capacitors are used and, as a result, the unit of farads is often very small. The units used for measuring are:

- microfarad μF (one millionth of a farad or 1/1,000,000);
- nanofarad nF (one thousandth of a microfarad or 1/1,000,000,000);
- picofarad pF (one thousandth of a nanofarad or 1/1,000,000,000,000).

Many types of capacitors exist but the most common types are polarised (tantalum and electrolytic) and non-polarised (polyester, mica, ceramic and polystyrene).

An electrolytic capacitor has a larger capacitance value, meaning it can store larger amounts of charge. The capacitor is also polarised, meaning it has a positive and negative pole and it must be connected into the circuit correctly. The casing of the capacitor is marked with its value and also a '−' sign, indicating the negative leg. If a capacitor is connected incorrectly, it will be permanently damaged.

Tantalum capacitors are often smaller in size than electrolytic capacitors and this often results in their use in smaller, compact circuits.

Non-polarised capacitors have smaller capacitance values and are, therefore, much smaller in size. They are also safer to use in an electrical circuit as they can be connected any way round. There are many types of non-polarised capacitors, such as polyester, mica, ceramic and polystyrene. Each capacitor is named after the material it is made from.

## Activity

Electronic components come in various shapes and sizes and it is always a good idea to be familiar with the most common forms. In small groups, select a component to research. Your task is to produce a simple handout showing the most common types of your component available. The handout should be clearly labelled and can be used as a visual aid until your familiarity with the components increases.

## Logic gates

Logic gates are digital devices commonly grouped together and found in integrated circuits (ICs). The logic gate monitors input signals, which can be either on (+ V) or off (0 V) and then, depending on its function, switches an output on or off.

All of the components described above are often referred to as discrete; this means they are individual components that are combined to form an electrical circuit. ICs are complete circuits constructed using many components, which are housed within a black, plastic casing. The IC is commonly referred to as a chip.

### AND gate

An AND gate with two inputs (A, B) would need both inputs A *and* B to be on for the output to be on.

### OR Gate

An OR gate with two inputs (A, B) would need input A *or* B to be on for the output to be on.

### NOT Gate

A NOT gate with one input (A) would give an output when input A is *not* on.

## Activity

Much more information is available about logic gates as they are a very common digital device. Using the internet or library, look up both the American and the British symbols of the three common logic gates. Look up the term 'truth table' and produce one of these for each device. Explain your findings to your teacher.

Once you have completed this research activity, create a logic gates quiz: each member of your class should submit a question to the teacher to form part of the quiz.

## Make the grade

The aim of this Make the grade is to allow you to demonstrate your understanding of six basic electronic components.

The next activity will help you in achieving the following grading criterion:

**P3**  describe the purpose of six different types of electronic component.

## Activity

**Clearly describe the purpose of the six different electronic components listed below.**

1. **Diode**
2. **LED**
3. **Transistor**
4. **LDR**
5. **Capacitor**
6. **Fixed resistor**

# Circuit diagrams

**P4**     **M1**

Now that we understand the operation and use of some of the most common components, it is time to start using them to produce what is called a circuit diagram. Explained simply, a circuit diagram is a drawing of an electronic circuit that shows all of the electronic components used as symbols. Using standard symbols allows the electronics engineer to figure out how the circuit works and also identify each component's value and location in the circuit.

A collection of standard electronic symbols is shown in Figure 19.11.

- In the circuit shown in Figure 19.12, standard symbols have been used to represent the components and connections needed to build a light-detecting circuit. The circuit works by using an LDR to monitor light conditions – during daylight, the LED will be off, but when a lack of light is detected (darkness), the LED switches on.

- In order to improve the understanding of the circuit diagram, some simple rules have been followed. The power supply or input into the circuit is typically shown

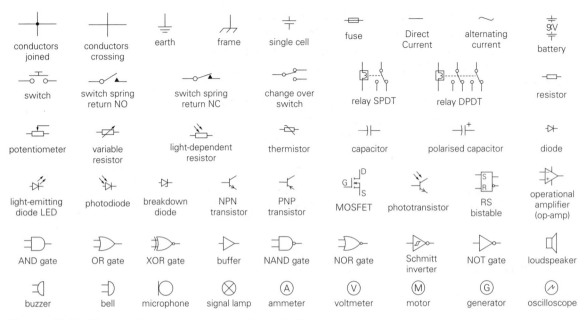

| | | | | | | | | | |
|---|---|---|---|---|---|---|---|---|---|
| conductors joined | conductors crossing | earth | frame | single cell | fuse | Direct Current | alternating current | battery | |
| switch | switch spring return NO | switch spring return NC | change over switch | relay SPDT | | relay DPDT | | resistor | |
| potentiometer | variable resistor | light-dependent resistor | thermistor | capacitor | | polarised capacitor | | diode | |
| light-emitting diode LED | photodiode | breakdown diode | NPN transistor | PNP transistor | MOSFET | phototransistor | RS bistable | operational amplifier (op-amp) | |
| AND gate | OR gate | XOR gate | buffer | NAND gate | NOR gate | Schmitt inverter | NOT gate | loudspeaker | |
| buzzer | bell | microphone | signal lamp | ammeter | voltmeter | motor | generator | oscilloscope | |

**Figure 19.11** Electronic components and symbols

on the left-hand side, and the output shown on the right-hand side. For this circuit, this is a 9V supply on the left and an LED on the right. The top rail is also the supply (or positive voltage) and the bottom rail is the common, ground or 0 V.

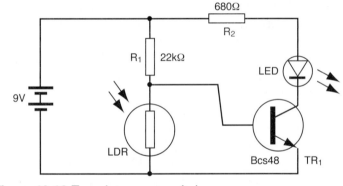

**Figure 19.12** Transistor as a switch

- In order to make diagrams look less complicated and avoid crossing too many wires, components can also be rotated and positioned in convenient places. For example, in Figure 19.12, R2 has been drawn vertically to save space.

- It is also common practice to use a solid dot to indicate the joining of tracks. This prevents the engineer getting confused when designing the circuit layout or building the circuit.

- The final piece of information shown on the circuit diagram is the component values. It can be seen that each component is clearly identified and an appropriate value is given. In the case of the transistor, this information is in the form of the transistor manufacturer part number.

A block diagram can also be used to represent an

electrical system and it is often the first stage of the design process. A block diagram for the transistor-switch circuit is shown in Figure 19.13. The input to the system is the LDR, which is used to detect light levels. The process element of the system is a transistor, which acts as a switch, turning on when the appropriate amount of light level is reached. The final element is the output; in our circuit, this is an LED, which turns on to indicate darkness.

**Figure 19.13** Block diagram for the transistor-switch circuit

## Circuit operation

We can now use the circuit diagram to determine the operation of the circuit. The 9V battery is used to power the circuit; it provides 0 V to the bottom rail and 9V to the top. Resistor R1 and the LDR are arranged to provide what is referred to as a potential divider. This is used to provide a specific output voltage to the transistor, making it turn on when the right conditions are met. In daylight conditions, the LDR has a very low resistance of around 400 $\Omega$. This means that when the circuit supply is divided, the LDR has a very low voltage, which is not enough to turn the transistor on. In the dark, the resistance of the LDR increases and it takes a large share of the voltage. This then turns the transistor on. R1 is also used to protect the base of the transistor from too much current.

With the transistor turned on, current flows through R2, which acts as a voltage limiter protecting the LED. The LED then lights up as the cathode connection is connected to 0V through the transistor.

## Resistor colour coding

Looking at the circuit diagram in Figure 19.12, you can see that resistors come in various values. It would be impractical for manufacturers to make every value possible, so a range of resistors are available to choose from. To tell us the value of a resistor, the manufacturer marks the resistor with a series of bands, each band representing a number. Figure 19.14 shows how to work out the resistance value.

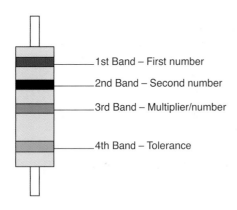

| Colour | Value |
|--------|-------|
| Black | 0 |
| Brown | 1 |
| Red | 2 |
| Orange | 3 |
| Yellow | 4 |
| Green | 5 |
| Blue | 6 |
| Violet | 7 |
| Grey | 8 |
| White | 9 |

| Tolerance colour | +/− |
|------------------|-----|
| Brown | 1 |
| Red | 2 |
| Gold | 5 |
| Solver | 10 |
| None | 20 |

1st Band – First number
2nd Band – Second number
3rd Band – Multiplier/number
4th Band – Tolerance

**Figure 19.14** Resistor colour coding

We can calculate the value of the resistor in Figure 19.14 in the following way:

- Band 1 = brown = 1
- Band 2 = black = 0
- Band 3 = orange = $10^3$ or 000
- Band 4 = gold = +/− 5

Value = 10,000 Ω or 10kΩ +/− 5 %

Letters can also be used to represent the value of ohms – R replaces Ω, k replaces kΩ and M replaces MΩ.

## Activity

Work out the value or colour code of the following resistors.

1. **Red, red, red gold**
2. **Yellow, red, blue, silver**
3. **10 Ω**
4. **2M7**

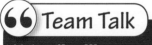 **Team Talk**

Aisha: **'I will never remember that colour code. Is there an easier way?'**
Steve: **'You can always remember the rhyme 'Bobby Brown Rides Over Your Grass But Violet Grey Won't' to give you the colours, then just add 0 to 9.'**

## Make the grade

The aim of this Make the grade is to allow you to demonstrate your understanding of basic electronic components. You must be able to apply this knowledge and identify the electronic components used within an electronics diagram and their function.
The next activity will help you in achieving the following grading criteria:

**P4** read a given circuit diagram to indentify the electronic components in the circuit

**M1** explain the function and operation for different electronic components

## Activity

1. **With reference to the burglar alarm transistor circuit shown in Figure 19.15 opposite, clearly identify all of the components shown in the diagram, stating their value or part number as appropriate.**

**Figure 19.15** Burglar alarm transistor circuit

2. **Clearly describe the function in the circuit of four components you have identified in Question 1.**

# Learning Outcome 3. Know about the manufacture of electronic circuit boards

## Types of circuit board

**P5** **M2**

Once the electronic circuit has been designed, the next step is to produce a working circuit. To do this, the components must be mounted to a circuit board using the most suitable construction technique. Three main types of circuit board exist and the electronics engineer must choose the type that best meets their design requirements. Their decision could be based on many things and the features, advantages and disadvantages of the three main types of board will now be explained.

## Proto boards (bread boards)

'Proto boards' or prototype boards (to give them their full name) are also referred to as 'bread boards'. This type of board is perfect for testing an electrical circuit before it is permanently made. The reason for this is that the board is made up of special push-fit terminals that are linked horizontally or vertically. Therefore, the components can simply be pushed into place and electrical connection is made. Components can be easily replaced and the circuit can also be modified without the need for any rewiring. The only real disadvantages of proto boards are that they are intended for temporary use and small circuits, but they can be coupled together.

**Figure 19.16** Proto board

## Strip boards

Strip board is sometimes called Veroboard®; this comes from the Vero electronics company that invented them. This type of board is ideal for construction of one-off electrical circuits quickly and easily. It is supplied pre-drilled so that components can be mounted onto the surface. The legs are then soldered to copper tracks running in one direction underneath. The tracks are then broken as required to link the electrical components.
The main disadvantage of this board is that it can be difficult to transfer a circuit layout to a strip-board layout. If constructed poorly, many links and breaks are needed and the circuit can look untidy. Soldering can also be tricky and tracks can become linked by stray pieces of solder, causing short circuits.

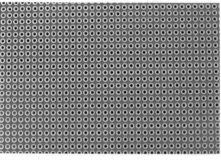

**Figure 19.17** Strip board

## Printed circuit boards (PCBs)

Printed circuit boards (PCBs) differ from the previous types of board as they are used in the batch production of electronic circuits. The PCB is made from either a strong card (in the case of cheaper PCBs) or, more commonly, a composite such as glass-reinforced plastic. One side of the board is plain for mounting the components and the other side is coated with a copper film. During the production process, some copper film is removed to leave the copper 'tracks', which act as wires to connect the components together and form the circuit. All that remains is for the PCB to be drilled (or pressed) out and components mounted and connected.

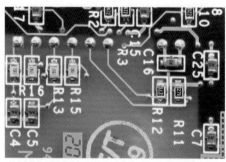

**Figure 19.18** Printed circuit board

The PCB has many advantages – it is very fast to produce, and it is reliable and cheap. It can also be very compact, with components mounted on both sides of the board. New designs use components that are surface-mounted to both sides of the PCB, and by coating the holes with copper, they can act as tracks between the two.

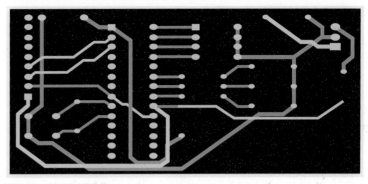

**Figure 19.19** PCB mask

The main disadvantage of using a PCB is that the initial set-up costs for a company will be very high. Specialist CAD and CAM software, CNC robotics and machine tools are required, as are the associated specialist skills, just to produce a prototype board. The PCB will then go through extensive testing before being placed into a device.

## Activity

There is a great deal of interesting information on the internet about PCBs. Why don't you try typing 'PCB design' and 'printed circuit board' into Google and see what you find. Use the internet to answer the following questions.

1. In what electrical item did the first PCB appear?
2. When was the first PCB produced?
3. When did surface-mount technology emerge?

### Printed circuit board design and construction

Like many tasks, PCBs can be designed and constructed by hand or by using an automated process. The following steps are used to produce a working PCB design. At each stage, the hand and auto techniques will be examined.

1. Stage 1: PCB layout diagram
   During this stage, the electronic circuit will be transferred to a PCB layout. By hand, this will involve moving the components around so that the track layout is simple and very few link wires are needed. Design techniques involve placing components such as resistors in a parallel position and also running track beneath ICs to save space. Components should be evenly spaced with tracks not too close together. The automated layout design process involves using a CAD package to automatically transfer the circuit diagram to a PCB layout diagram. The software will consider the points explained above and produce the most practical layout available.

2.  Stage 2: Producing the artwork mask
    The mask is the name given to the diagram that shows the track and pad layout of the PCB. Figure 19.19 shows a typical PCB mask. To produce the mask by hand, a piece of tracing paper can be placed over the layout diagram produced in Stage 1. The layout of the tracks and circular pads are then traced onto the paper. To produce a mask for a CAD-designed layout, it is simply a matter of printing the design onto a thin acetate sheet.

3.  Stage 3: Transferring the mask to the board
    When making the PCB by hand, the copper board should first be cut to size and then the copper side cleaned using a scourer. Once cleaned, the copper side should be kept clean from dirt. The tracing-paper mask is turned upside down so that it can be copied to the clean copper side of the board; doing this will ensure the layout is correct when mounting components on the top of the board. A thin-tipped waterproof permanent marker is now used to draw the pads and tracks onto the copper. The procedure for transferring the acetate mask is quite different; the copper on the board is coated with a photosensitive material, which is protected by a polymer film. Away from direct sunlight, the film is removed and the mask is placed onto the board, making sure the printed side is touching the copper. The board and mask are then placed into a UV exposure box, which transfers the layout onto the photosensitive material. Once the exposure is complete, the board is placed into a developing solution, using appropriate PPE, for about ten seconds. This exposes the PCB layout, and the board should be thoroughly washed under cold running water.

4.  Stage 4: Etching
    The etching process can be carried out in large etching tanks for mass-produced boards or in small containers for hand-produced PCBs. The etching process involves bathing the copper board in a highly corrosive etching solution to remove the unwanted copper. Once complete, only the copper pads and tracks will remain on the PCB. The etching time varies with each design and there is a risk of damaging the copper track if the board is left in the solution for too long.

5.  Stage 5: Cleaning
    The PCB is now cleaned to remove any etching solution and the permanent marker or photosensitive film is removed using a PCB eraser or abrasive pad. Removing the film will improve the soldering process and allow the tracks to be checked for any slits.

6. Stage 6: Drilling

This stage of the process involves drilling the holes to mount the components. Holes should be drilled from the track side to ensure the pads and tracks are not pushed off the board as the bit breaks through. This process may involve using various drill bits appropriate to the component leg size. In automatic production, a CNC machine tool would typically be used to speed up the process and provide extreme accuracy.

7. Stage 7: Placing and soldering components onto a PCB

In the final stage of the process, all of the necessary components are added to the PCB. Extra care is taken to ensure that the correct pin layout is identified and that components are placed in their correct orientation. Resistor and capacitor values are double-checked and, once all components are positioned, soldering can begin. Once the soldering is complete, the track side of the PCB is often sprayed with a protective lacquer to protect it from dirt and surface damage.

A number of other techniques are also used to produce PCBs, including screen printing and pressing. These tend to be associated with cheaper PCBs where a card board is used rather than a composite.

## Make the grade

By completing this Make the grade, you will demonstrate an understating of the types of circuit boards used when constructing electronic circuits. The next activity will help you in achieving the following grading criteria:

**P5** describe the manufacture of the three types of electronic circuit boards;

**M2** explain the advantages of the three types of circuit board.

## Activity

1. **Clearly explain how a circuit would be manufactured using the following types of circuit boards:**
   a) **proto boards (bread boards);**
   b) **strip boards;**
   c) **printed circuit boards (PCBs).**
2. **Explain the advantages and disadvantages of the three types of circuit boards in Question 1.**

# Learning Outcome 4. Be able to construct an electronic circuit

## Circuit construction techniques

When constructing an electronic circuit, several techniques are available to make the process easier and to improve the circuit's performance. Whether or not these techniques are used depends on many factors; these factors could be cost, time, resources available and also the intended use of the circuit. Eight techniques will now be examined.

## Wire wrapping

This construction technique was very popular during the 1960s. Wire-wrapped circuits appeared in everything from radios to space-shuttle guidance computers, due to their reliability and the ease with which they can be modified. The method of wire wrapping was quite simple. Small electronic assemblies would be built, mounted to square metal posts. These posts would then be linked as required by wrapping a link wire between each post. This process could be completed manually or fully automated.

## Soldering

Soldered connections can be found on PCBs and strip boards as the main joining method. The process involves using a hot soldering iron to heat the two conductors that need joining and then fusing them together using a soft solder, which melts at about 200 degrees Celsius, joining the two pieces together when the solder cools. Soldering produces connections that have little resistance and, therefore, it is as good as wiring but it also provides a mechanical joint of good strength. Soldering can also be fully automated in the production of a PCB, using a paste that is printed onto the PCB before the components are laid onto the circuit (in the case of surface-mount components) and using solder baths (in the case of 'through-hole' assemblies)

## Heat shunt

When soldering, a heat shunt can be used to divert or shunt heat away from delicate components.

## Heat sinks

It is quite common for electrical components to produce heat. Unfortunately, if the heat becomes too great, a component can be destroyed. One solution to this is to introduce cooling in the form of a heat sink. A typical heat sink is made from an aluminium alloy and has one flat surface, which is attached to the component, and then a series of combs or fins, which are used to dissipate the heat while being cooled by air. Heat sinks are often found in power-supply circuits.

## Surface-mount technology

Surface-mount technology (SMT) is a manufacturing technique used to populate PCBs, with the circuit board components mounted directly onto the surface of the PCB. The incentive for this type of technology has been consumer need for increasingly more compact and portable devices, and the manufacturers' need for reduced manufacturing costs with greater performance and reliability.

So how does SMT differ from 'through-hole' technology? We can start by looking at the size of the components or surface-mount devices (SMDs). The smallest component available is the resistor. The '0201 package' measures a tiny 0.6 mm × 0.3 mm. Components are typically a flat cylindrical shape with contacts either end in the form of flat pads, short pins or a matrix of soldered balls.

**Figure 19.20** Surface-mount PCB

Apart from the fact that no holes are needed for mounting components, the SMT circuit board also differs in its design. Tracks are generally thinner and shorter, with a reduced number of wire links. The circuit boards are manufactured with the circuit design, and pads are placed where the components are to be attached. These pads are then pre-soldered ready for component termination. Components are then placed onto the board by an automated process before the PCB passes into a reflow soldering oven. The oven has three zones. The first zone preheats the board and components gradually to a desired temperature and they then pass into a flowing zone. In the flowing zone, the temperature is raised to a level where the solder on the pads begins to melt. The component legs are then bonded to the board and the PCB finally passes into a cooling zone.

A completed PCB using surface-mount technology is shown in Figure 19.20.

Looking at Figure 19.20, we get an idea of how small the components are. It goes without saying that the resistors are too small to be marked with their colour codes. So that their resistance can still be identified, they are marked with three digits. The first two numbers are the first two numbers of its value and the third number indicates how many zeros. Figure 19.20 shows a resistor marked 222; the value of this would be 2,200 Ω or 2.2 kΩ.

## Double-layer board

So far we have only looked at single-sided circuit boards. However, there is the option of using double-sided circuit boards. As the name suggests, this type of board has components and tracks on both of its surfaces. While this makes its design much more complicated using through-hole technology, it does reduce the size and cost of the circuit board. With surface-mount components, it is far easier to use a double-sided board because no through holes are required.

Double-sided circuit boards can also be used to combine technologies; it is not uncommon to have surface-mount components on one side of the board and through-hole components on the other.

## Component pin identification

Components such as transistors and ICs have connecting pins, which require correct termination into a circuit. If these pins are not identified and terminated correctly, then damage may occur to the circuit or component.

If we look at the circuit symbol for a transistor, we know that it has three terminals: a base, collector and emitter. There are several configurations for the pins, which depend on the transistor casing. Figure 19.21 shows the arrangement of some of the most common types of transistors when viewed from the underside.

A typical IC is the NE555N. The timer is housed in an eight-pin dual-in-line (DIL) package, which means there are two parallel rows of four pins. Each of these pins has a specific role and it is vital that they are correctly identified. You can see from Figure 19.22 that there are two markers on the casing of the 555 timer IC to indicate its orientation.

The semi-circle indent is used to indicate the top of the IC and the circular indent is used to indicate Pin 1. Pins 2 to 8 then follow in an anti-clockwise direction.

## Power supply connections

A power supply unit (PSU) is often used as an alternative to a battery to power low-current electronic equipment and circuits. PSUs can have fixed or variable voltage outputs allowing one PSU to have the flexibility to supply various circuits. Fitting a power supply connection to a circuit rather than using a battery means there is no risk of losing power. It also ensures a constant voltage value is applied to a circuit, unlike a battery which will reduce as the circuit is in use.

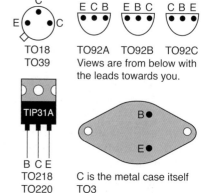

**Figure 19.21** Transistor pins

**Figure 19.22** 555 timer

## Noise

Electronic noise is generally an unwanted signal that becomes superimposed on the signal being processed. Electromagnetic interference (EMI) is a type of noise found in electronic circuits containing magnetic fields. The effects of EMI can be reduced by using cable that is protected or screened with a layer of interwoven copper strands. This screening would then be connected to the circuit at 0 volts, or ground.

### Activity

A circuit board manufacturing company is to produce the transistor burglar alarm circuit shown in Figure 19.15. The circuit must be cheap to manufacture, reliable and compact. Propose a suitable method that could be used and fully justify your choice.

### Make the grade

By completing this Make the grade, you will demonstrate an understating of circuit construction techniques. The next activity will help you in achieving the following grading criterion:

**D1** propose a method used to construct a given electronic circuit and justify your choice.

## Types of circuit

P6   M1

The transistor is a very popular component when designing simple electronic circuits. The use of the transistor as a switch has previously been examined and, in the circuit in Figure 19.23, it is used again to switch on a burglar alarm circuit.

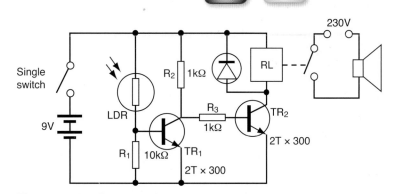

**Figure 19.23** Burglar alarm circuit

This circuit operates on the same principle as the light-detector circuit, except that its action is reversed. When the circuit is switched on and if light from a torch or lamp shines on the LDR, both TR1 and TR2 are turned on, causing the relay to latch in. The relay contact then closes to supply a mains voltage of 230V to the siren. This circuit also contains a diode to protect the transistor from a brief voltage surge, which occurs when the relay turns off. R1 is also a variable resistor, which allows the sensitivity of the LDR to be adjusted.

A single-stage transistor amplifier is shown in Figure 19.24. This is another use of the transistor.

This circuit would take small input signals from a sensor or microphone and produce a larger signal at the output. The increase in signal size is called the gain of the amplifier. The amplifier gain with the biasing resistors shown should be 100 for the output.

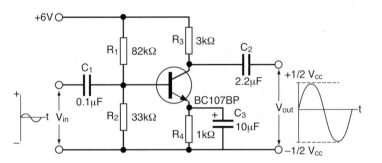
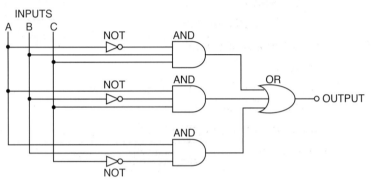

**Figure 19.24** Single-stage transistor amplifier

The final type of circuit to be examined is a logic circuit. The combinational circuit shown in Figure 19.25 has a total of six logic gates, which work together to produce an output only when the correct sequence of inputs is supplied.

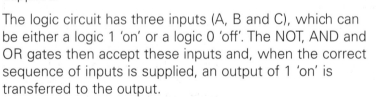

**Figure 19.25** Combination logic circuit

The logic circuit has three inputs (A, B and C), which can be either a logic 1 'on' or a logic 0 'off'. The NOT, AND and OR gates then accept these inputs and, when the correct sequence of inputs is supplied, an output of 1 'on' is transferred to the output.

Logic gates are often combined into ICs, with typically four identical gates per chip. These chips are then mounted to PCBs with a track layout designed to connect the gates in the necessary logic combination.

## Make the grade

By completing this Make the grade, you will demonstrate an understating of the methods used to construct a given electronic circuit safely in the laboratory. Consideration should also be given to the hazards associated with this activity.
The next activity will help you in achieving the following grading criteria:

**P2** use safe working practices in the electronics workshop/laboratory

**P6** use two methods of construction for a given simple electronic circuit;

# Activity

1. You are to build the transistor as a switch circuit shown in Figure 19.26 using two methods of construction.

    a) The first method of construction is to be proto board. This will allow the circuit components to be easily tested and the circuit operation to be verified.

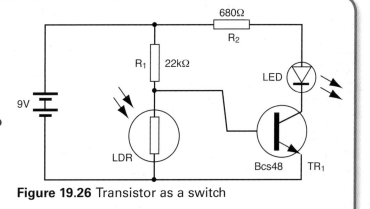

**Figure 19.26** Transistor as a switch

    b) The second method of construction is to be strip board. This method is ideally suited to our one-off design.

2. Obtain a witness statement from your tutor to prove safe working practices were used during the construction of the circuits.

## (r) Grading criteria recap

To achieve a pass grade you must be able to:

**P1** describe the potential hazards related to constructung electronic circuits

**P2** use safe working practices in the electronics workshop/laboratory

**P3** decribe the purpose of six different types of electronic component

**P4** read a given circuit diagram to identify the electronic components in the circuit

**P5** decribe the manufacture of the three types of electronic circuit boards

**P6** use two methods of construction for a given electronic circuit

To achieve a merit grade you must be able to:

**M1** explain the function and operation of four different electronic components

**M2** explain the advantages and disadvantages of the three types of electronic board

To achieve a distinction grade you must be able to:

**D1** propose a method used to construct a given electronic circuit and justify your choice.

# Activity answers

## Unit 3

### Activity Answers, page 90
1. 161910
2. 3116
3. 24
4. 1184944

## Unit 4

### Activity Answers, page 155

Distance – Time graph

1. A–B: $\dfrac{10 \text{ metres}}{2 \text{ seconds}} = 5 \text{ m/s}$

2. B–C: $\dfrac{0 \text{ metres}}{5 \text{ seconds}} = 0 \text{ m/s}$

3. C–D: $\dfrac{5 \text{ metres}}{3 \text{ seconds}} = 1.67 \text{ m/s}$

4. D–C: $\dfrac{0 \text{ metres}}{20 \text{ seconds}} = 0 \text{ m/s}$

Velocity–Time graph

1. C–D: the car accelerated from 8 m/s to 12 m/s in 9 seconds
   Acceleration $= \dfrac{4 \text{ m}}{\text{s/9}} = 0.44 \text{ m/s}^2$

2. D–E: the car decelerated from 12 m/s to 0 m/s in 10 seconds
   Deceleration $= \dfrac{12 \text{ m}}{\text{s/10}} = 1.2 \text{ m/s}^2$

### Activity Answers, page 156

$A = \dfrac{1}{2} \times b \times h = 0.5 \times 3 \times 6 = 9$ metres

$B = b \times h = 5 \times 6 = 30$ metres

$C = b \times h = 4 \times 8 = 32$ metres

$D = b \times h = 9 \times 8 = 72$ metres

$E = \frac{1}{2} \times b \times h = 0.5 \times 9 \times 4 = 18$ metres

$F = \frac{1}{2} \times b \times h = 0.5 \times 10 \times 12 = 60$ metres

Total distance = 221 metres

# Activity Answers, page 159

Newton's laws of motion:

1. Every object will remain in a state of motion or rest unless an external force is applied.
   *This is why objects accelerate (when a force is applied) and why they decelerate (due to forces opposing motion, such as friction).*
   The relationship between an object's force, mass and acceleration is F = m x a.
   *This is how we calculate the force of an accelerating object or how we calculate the weight of an object from its mass.*
   Every action has an equal and opposite reaction.
   *This is how we can determine static forces and how we can determine the normal force, and hence the coefficient of friction.*

2. To answer this question (and similar questions), you need to remember that:

$$\text{Kinetic coefficient of friction} = \frac{\text{kinetic friction force}}{\text{normal force}}$$

The normal force is equal and opposite to the weight, so it can be calculated by the given mass.

Normal force = mass × gravity = 3 × 9.81 = 29.43 N

From this, we have the coefficient of kinetic friction and the normal force; we can now calculate the kinetic friction force by rearranging the formula.

Kinetic friction force = kinetic coefficient of friction × normal force
Kinetic friction force = 0.23 × 29.43
Kinetic friction force = 6.7689 N

We also need to calculate how far the steel block has travelled. If it has a velocity of 0.75 m/s then it will travel 0.75 metres in 1 second. In 1.3 seconds, it will travel:

Distance travelled = velocity × time
Distance travelled = 0.75 × 1.3
Distance travelled = 0.975 m

Once this is completed, it is relatively straightforward to calculate the work and power using the formula shown below:

Work done = force (frictional kinetic) × distance travelled
Work done = 6.7689 × 0.975
Work done = 6.6 J

To complete the question:

Power = work done / time taken
Power = 6.6 / 1.3
Power = 5.08 W

## Activity Answers, page 163

$$\text{Current} = \frac{\text{change}}{\text{time}}$$

$$I = \frac{Q}{t}$$

## Activity Answers, page 164

a. $I = \frac{V}{R}$

b. $R = \frac{V}{I}$

# Unit 8

## Activity Answers, page 190

A key part to the answer is to identify both steels as completely different materials with different properties, even though they are both steel.

Low-carbon steel is ductile so can be easily shaped; it is also tough so provides good protection. However, it would be too soft for a chisel, so the cutting edge would become blunt very quickly. Further investigation may include heat treatment, whereby the chisel (medium-carbon steel) could be tempered to reduce hardness and increase toughness; this would reduce the chance of damage. This would not be possible with low-carbon steel as it reacts differently to heat treatment.

## Teacher Guidance for D1 (Activity, page 217)

Different materials can be used for this answer (e.g. mild steel or aluminium) so the student has to justify their choice. They should show a good appreciation of the properties in the question but should also extend their answer with other properties that have been discussed in this unit. It is a Distinction question so additional information, such as results from formal or informal testing of materials, will be an advantage, as well as research into the manufacture of the given application.

# Unit 14

## Teacher Guidance for P1 (Activity, page 269)

Grading criterion P1 is assessed by a piece of written work and the student should show that they can distinguish between the operation of different machines and what the machines are capable of producing. The criterion is restricted to 'describing' and only the basic understanding of the machine is required. One hundred and fifty words to describe each machine will be enough to show basic understanding.

## Teacher Guidance for D1 (Activity, page 269)

For D1, the student is expected to justify a choice of secondary machining for a component. The component does not necessarily need to be complicated. The example shown has only a few features but its method of manufacture is not immediately obvious. It could be produced partly by a centre lathe, partly by a vertical milling machine or wholly by a CNC milling machine. The student should be able to compare different methods based upon experience and then justify their choice of manufacture. Therefore, it is possible for different students to give different methods of manufacture. It is the justification that is the important part of this criterion.

## Teacher Guidance for P2 (Activity, page 275)

Grading criterion P2 is assessed by a piece of written work and the student should show that they clearly understand where a work-holding device should be used. The grading criterion asks for three different devices. Students should use three devices from three different machines to show breadth of knowledge. Around 100 words to describe each work-holding device will cover this criterion.

## Teacher Guidance for P3 (Activity, page 282)

Grading criterion P3 is assessed by a piece of written work and the student should show that they clearly understand the use of tools for a range of machining techniques. The grading criterion asks for three different tools. Students should use three tools from three different machines to show breadth of knowledge. Around 100 words to describe each tool will cover this criterion.

## Teacher Guidance for P4 (Activity, page 288)

Grading criterion P4 is assessed by practical application of the use of a machining technique by the student. The student must be considered to be able to operate the machine safely before this assessment can start. The grading criterion indicates that one workpiece is to be produced. For simplicity, a single machining process such as milling or grinding could be used. However, a product that uses two processes could be used. A witness statement could be used as written evidence.

The student should set speeds and feeds, using calculation or tables. These are recorded in the table on page 384.

| Setting machine parameters | | | | | Witness signatures | |
|---|---|---|---|---|---|---|
| **Select and datum edges** _Bottom left corner_ | | | | | _Signature_ | |
| **Set datum edges** _Use edge finder on left edge and front edge._ _Touch cutting tool on top of workpiece for height._ | | | | | _Signature_ | |
| | | | | | **Student sets machine** | |
| | Operation | Tool | Speed | Feed | Speed | Feed |
| 1 | Mill shoulder 11.5 mm deep | 20 mm End mill | 480 r.p.m. | 190 mm/ min | _Signature_ | _Signature_ |
| 2 | Mill slot 10 mm × 50 mm | 20 mm End mill | 950 r.p.m. | 300 mm/ min | _Signature_ | _Signature_ |
| 3 | Mill slot 10 mm × 50 mm | 10 mm HSS twist drill | 600 r.p.m. | 360 mm/ min | _Signature_ | _Signature_ |

Setting machine parameters

# Teacher Guidance for P5 (Activity, page 299)

Grading criterion P5 is assessed by practical application of the use of machining techniques by the student. The focus of the criterion is checking the machined component for accuracy and confirming that the student can use machinery safely. The teacher can record the students' performances using a witness statement.

# Teacher Guidance for M1 (Activity, page 300)

Grading criterion M1 builds upon criterion P5 by asking the students to explain the need for the checks that are carried out. The student should be able to explain that the checks during manufacture will prevent products that are incorrect having further work done to them. They should also consider that detecting errors may prevent further errors being made in subsequent products. They could consider internal customers inside the manufacturing company, who will reject components, and should also consider the end user, who will return products.

# Teacher Guidance for P6 (Activity, page 307)

Grading criterion P6 is assessed by practical application of the use of a machining technique by the student. The focus of the criterion is the understanding of machine safety. It can take the form of an essay or bullet points with detailed information, but it should rely on actual experience of the use of a machine as well as researching written information.

## Activity Answers, page 307

The five general methods could relate to PPE, guards or tidy work areas. The machine-specific methods of reducing risk could be as follows.

### Milling machines

Ensure that tools such as mallets or spanners are removed from the worktable while machining takes place. This will prevent them falling from the machine during operation. Double-check that tool-holding devices are securely holding the workpiece before machining starts. This will prevent the workpiece being forced out of position during machining. Do not try to remove swarf with a brush while the machine is in operation. The brush or a hand could get caught up in swarf. The sharp swarf could easily cut a hand.

## Teacher Guidance for M2 (Activity, page 307)

Grading criterion M2 requires the student to be more detailed in their knowledge and they should explain problems rather than just describe them.

## Teacher Guidance for D2 (Activity, page 308)

In order for the student to show a full understanding for grading criterion D2, a range of machines should be used. A pedestal drill, a milling machine or lathe and a grinding machine should be specified in the task. In this way, the student can contrast between the relatively inaccurate pedestal drill, the lathes with 0.01 accuracy, and then the grinding machines with 0.001 accuracy.

This will also give students the opportunity to show an understanding of a range of safety issues, from drills with no table movement but high speeds, to the dangers of revolving chucks on centre lathes and of wheels bursting on grinding wheels.

# Unit 18

## Teacher Guidance for Activity, page 310

Teacher's note: some equipment given to students should include common problems caused by workshop use.
The students should be able to detect any faults and give recommendations of how to correct the faults.

# Index